LearnCNC ™

for Haas Machine Tools

Advanced Mill CNC Programming and Applied Mathematics

This intermediate to expert level manual provides advanced application of codes, programming, and use of canned cycles. The content and sample programs provided cover a broad range of CNC programming requirements.

Advanced mathematics and formulas including applied shop floor trigonometry and geometry are used.

Learning materials may be used in conjunction with optional Virtual Training Environment for CNC including virtual CNC panels and 3D machines. Details provided in manual.

```
O405  ( ORIGIN LOWER-RIGHT)
G28 G91 G0 Z0
G54 G90 G0 X1.0 Y-1.0
T2 M6
S1000 M3
G43 H2 Z0.1 M8
G1 Z-0.1 F10.0
G41 D2 Y0
X-4.0
G2 X-4.3548 Y1.1395 I0 J0.625
G1 X-1.7636 Y2.9261
G2 X0 Y2.0 I0.6386 J-0.9261
G1 Y-1.0
G40 X1.0 (turn off CC)
G28 G91 G0 Z0
M30
```

AF394164

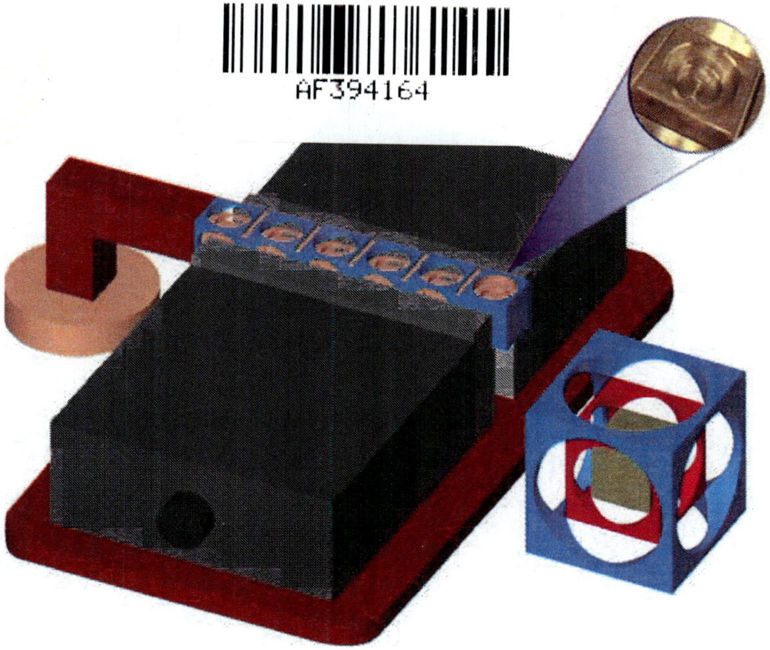

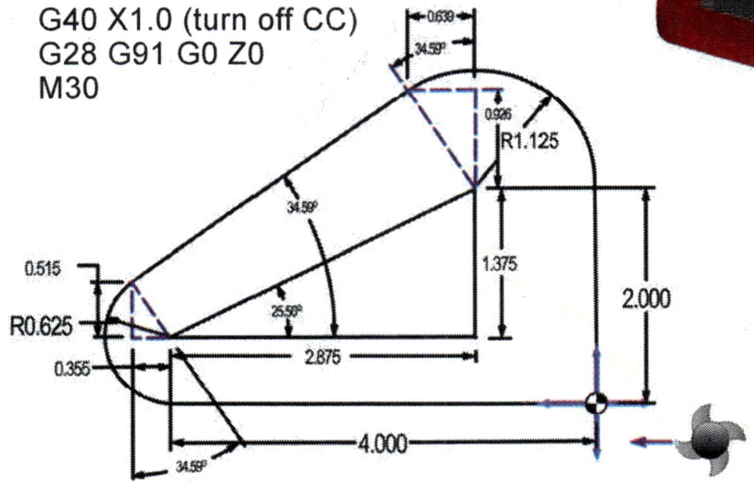

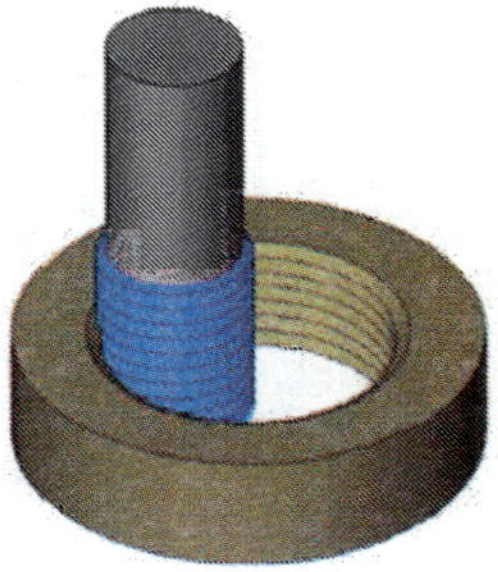

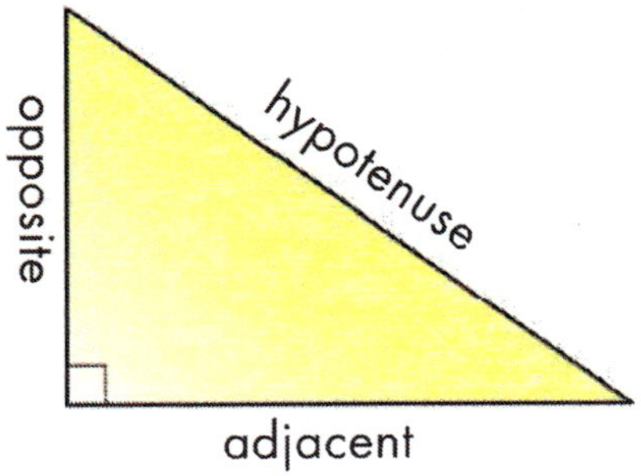

ISBN-13: 978-0-9816982-1-2

LEVEL 2

Powered by immerse2learn.com

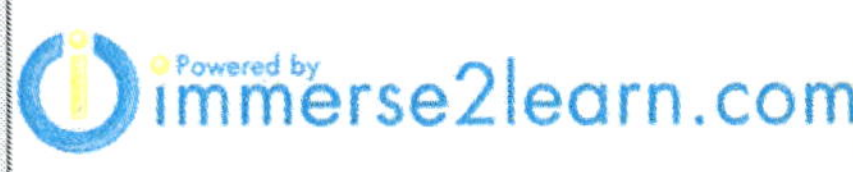

Learning Productivity Tools

Virtual Training Environment
CNC Machining

Flight Simulator Technology

The Virtual Training Environment for CNC Machining combines *powerful* "flight-simulator" technology with a *flexible* Internet-based learning content management system to deliver a truly innovative learning experience.

Simulated 3D machine models

Mimic Reality

- Virtual mills and lathes

- Industrial control panels

- Edit, load, run and save NC programs

- Set tool and work offsets

- Touch probes

- Stock material removal

- Canned cycles

- Control alarms

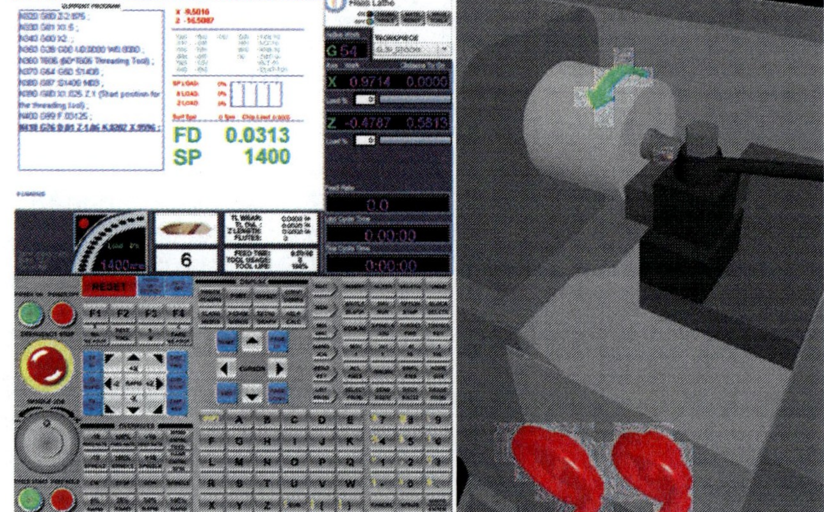

Real-time material removal

Mill and lathe touch probes

Learning Productivity Tool

Unlimited access to train and rehearse in the Virtual Training Environment for CNC machining enables learners to develop greater confidence and proficiency prior to performing actual procedures and operating equipment.

Learn at Your Convenience!

Learn CNC machine setup and programming with popular control panels for Haas Machine Tools.

Major Benefits

- Cost-effective and safe
- 24/7 access at your convenience
- Track and measure learner progress
- Reduce risk to people and equipment
- Increase control panel training contact-time
- Learn with virtual controls and 3D machines

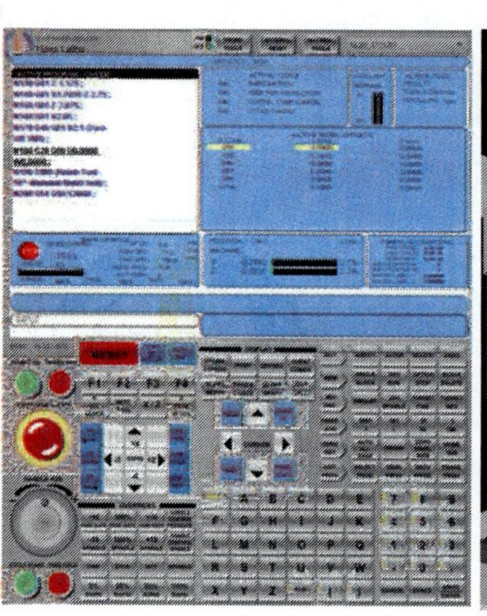

Popular control panels

Lathe applications

Additional Products

Online Learning Courses and Virtual 3D CNC Machine

Mill and *Lathe* CNC online courses provide the learner with comprehensive learning content, interactive exercises and virtual CNC panels and 3D machines. Depending on the course selected learning modules may include:

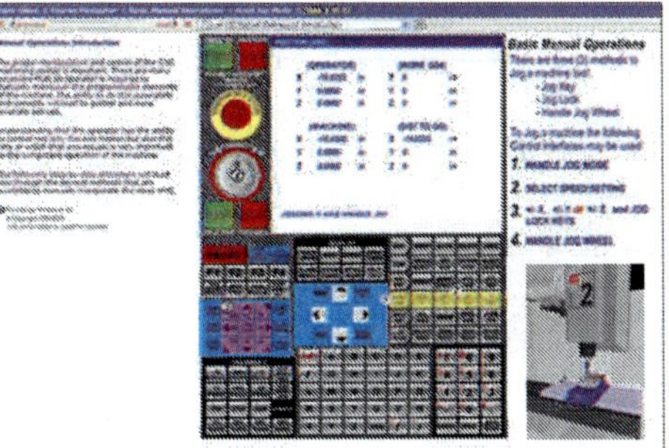

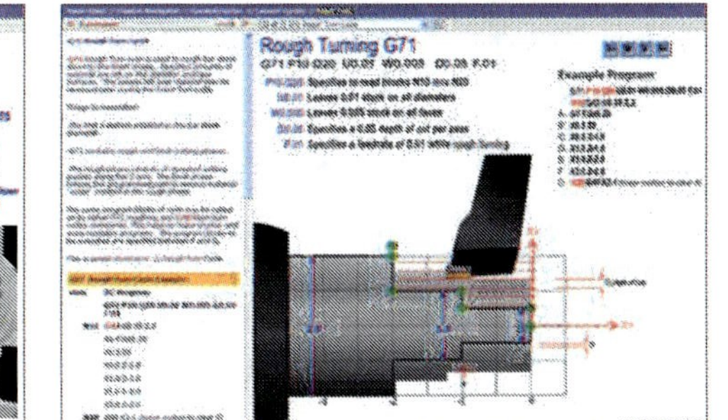

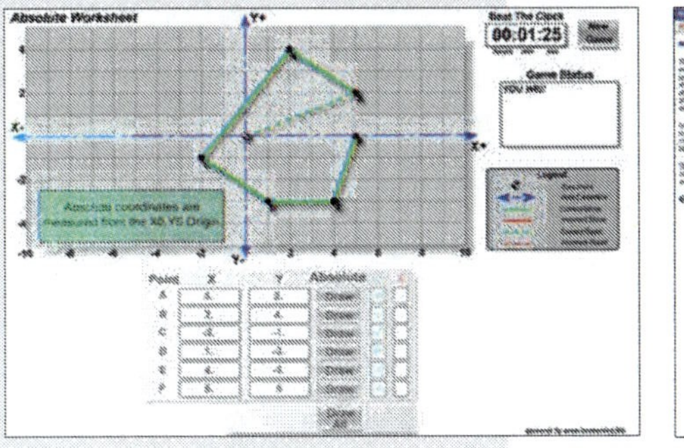

Interactive online courses

Setup

- Machine Motion Description
- CNC Panel Interface
- Machine Start-up
- Manual Operations
- Job Setup
- Edit Capabilities
- Program Entry
- Program Run

Programming

- Codes and Programs
- Program Structure
- Cartesian Coordinates System
- Cutter Compensation
- Tool Nose Radius Compensation
- Circular Interpolation
- Hole Manufacturing
- Programming Labs

Color Manuals, Exercises and Projects

Mill CNC Programming Level 1

An introduction to codes and programming, this manual is designed for beginner to intermediate level *Mill* CNC operators and programmers. The content and sample programs provided cover a broad range of CNC programming requirements. Basic mathematics and formulas are used.

Lathe CNC Programming Level 1

An introduction to codes and programming, this manual is designed for beginner to intermediate level *Lathe* CNC operators and programmers. The content and sample programs provided cover a broad range of CNC programming requirements. Basic mathematics and formulas are used.

Mill CNC Programming and Applied Mathematics Level 2

This intermediate to expert level manual provides much more advanced application of codes, programming, and use of canned cycles. The content and sample programs provided cover a broad range of CNC programming requirements. Advanced mathematics and formulas including applied shop floor trigonometry and geometry are used.

Other Products and Services

- Robotic simulation
- Learning content management system
- Online evaluation tools
- Content development
- Onsite training
- CNC programming

Table of Contents

Chapter 1
Subroutines & Subprograms

     Advanced CNC Mill Programming and Applied Mathematics Level 2

Objectives

1. The student will demonstrate knowledge of why subroutines and subprograms are used.

2. The student will demonstrate knowledge of the difference between subroutines and subprograms.

3. The student will program subroutines and subprograms.

4. The student will *loop* subroutines and subprograms.

     Advanced CNC Mill Programming and Applied Mathematics Level 2

Subroutines & Subprograms

Main programs have the ability to call **subroutines** and **subprograms**. Some of the benefits include: easy to read, ability to repeat a defined sequence multiple times, ability to translate and rotate, and simultaneous engineering.

Subroutines are *called* by **M97 P'n'**. The subroutine is located at the bottom of the program—*after* M30—preceded by **N'n'**.

   e.g., **M97 P100** calls a subroutine starting with N100 found after M30

Subprograms are *called* by **M98 P'n'**. Subprograms are stored as separate programs that look similar to main programs **O'n'**. They must be located in the memory of the control, along with the main program.

   e.g., **M98 P200** calls a CNC program named O200

Subroutines and subprograms return to the main program via **M99** at the end of the subroutine or subprogram.

Looping

Subroutines and subprograms can be executed (*looped*) multiple times by including **L'n'**, for Loop.

L specifies *Loop*.

'n' specifies how many times to loop.

Programmed wisely, looping provides an opportunity to program a segment of code and ask it to repeat several times while incrementing in Z between loops, or rotating around an axis, etc. Examples include gear cutting, profiling with multiple depth passes, etc.

M97 P100 L5 calls subroutine N100 and executes it 5 times.

M98 P150 L5 calls subprogram O150 and executes it 5 times.

SAMPLE SUBROUTINE

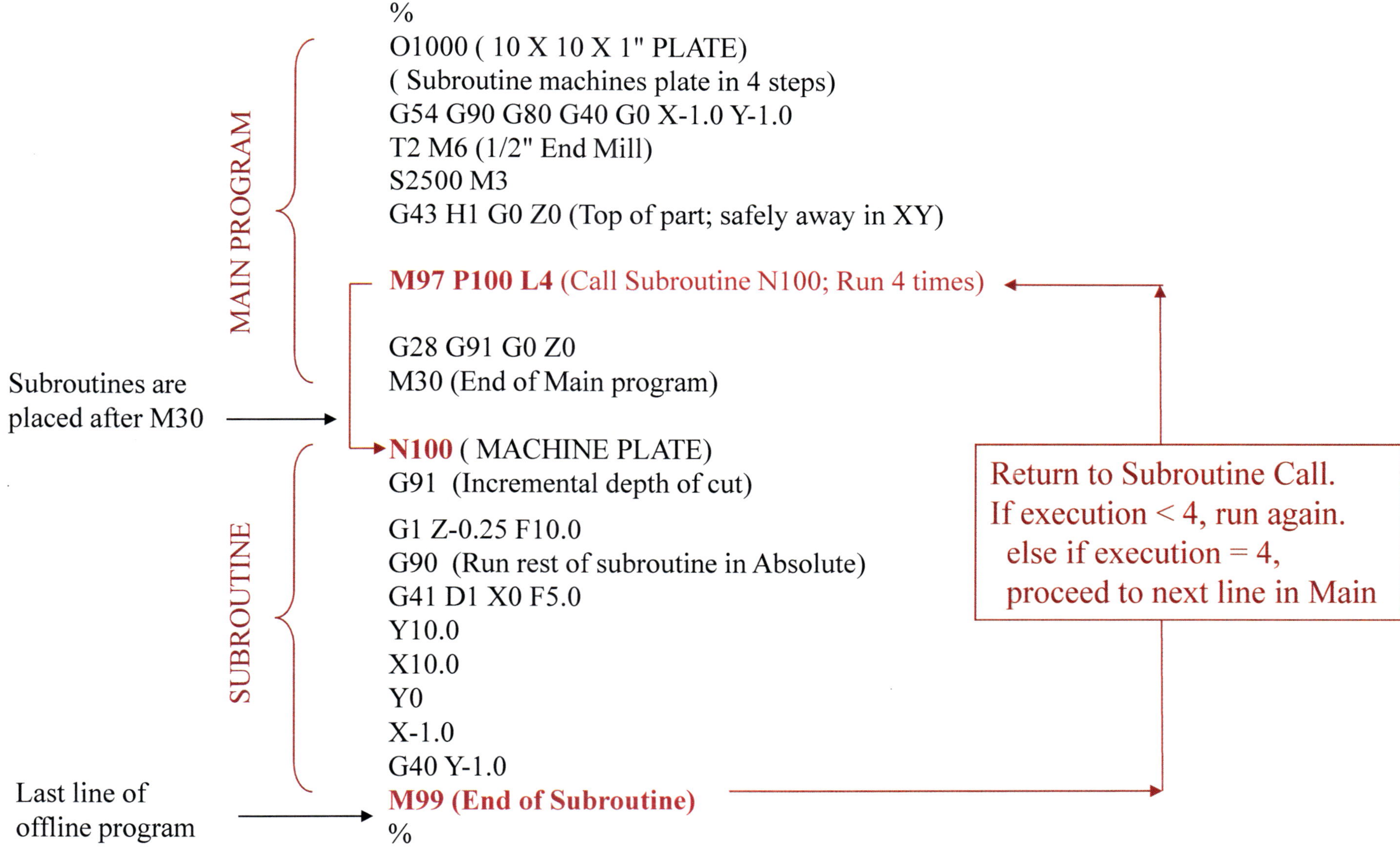

SAMPLE SUB PROGRAM

Main Program

```
%
O1000 ( 10 X 10 X 1" PLATE)
( Sub program machines plate in 4 steps)
G54 G90 G80 G40 G0 X-1.0 Y-1.0
T2 M6 (1/2" End Mill)
S2500 M3
G43 H1 G0 Z0 (Top of part; safely away in XY)


M98 P100 L4 (Machine plate in 4 steps)


G28 G91 G0 Z0
M30
%
```

Sub Program

```
%
O100 (Sub program for 10 x 10 x 1 Plate)
G91  (Incremental depth of cut)
G1 F10.0 Z-0.25
G90  (Run rest of subroutine in Absolute)
G41 D1 X0 F5.0
Y10.0
X10.0
Y0
X-1.0
G40 Y-1.0
M99 (End of Subroutine)
%
```

Return to Subprogram Call.
If execution < 4, run again.
  else if execution = 4,
  proceed to next line in Main

   Advanced CNC Mill Programming and Applied Mathematics Level 2

SAMPLE NESTING

Main Program

Subprogram

```
%
O1060
G54 G90 G0 X0 Y0
T1 M6  ( 3/8 END MILL  )
S8000 M3
G90 G0 X0 Y0
G43 H1 G0 G90 Z.1 M8
M98 P1050 L6  (6 Holes)

G28 G91 G0 Z0
M30
%
```

```
%
O1050 (CIRCULAR INTERPOLATE)

G91 G1 Z-.098   F80.0
G1 X-.0313 Y.0313
G3 X-.0313 Y-.0313 R.0313
M97 P100 L15 (Helical Interpolate 15x)
G3 I.0625 J0
G3 X.0313 Y-.0313 R.0313
G1 X.0313 Y.0313
G90 G0 Z.1
G91 X1.125 (SHIFT TO NEXT HOLE)

N100  (Subroutine)
G91 G3 I.0625 J0 Z-.02
M99 (END SUBROUTINE 100)

M99 (END SUBPROGRAM 1050 )
%
```

Subprograms can contain subroutines.
Each Subroutine contains N'n' and M99.
The subprogram looks like a main program
except it ends with M99.

Chapter 2
Circular Hole Milling

    Advanced CNC Mill Programming and Applied Mathematics Level 2

Objectives

1. The student will list the benefits of circular milling.

2. The student will program holes using circular milling with R and I J.

3. The student will helical interpolate.

4. The student will calculate the tool path using prescribed formulae containing the hole size and tool diameter.

Circular Hole Milling

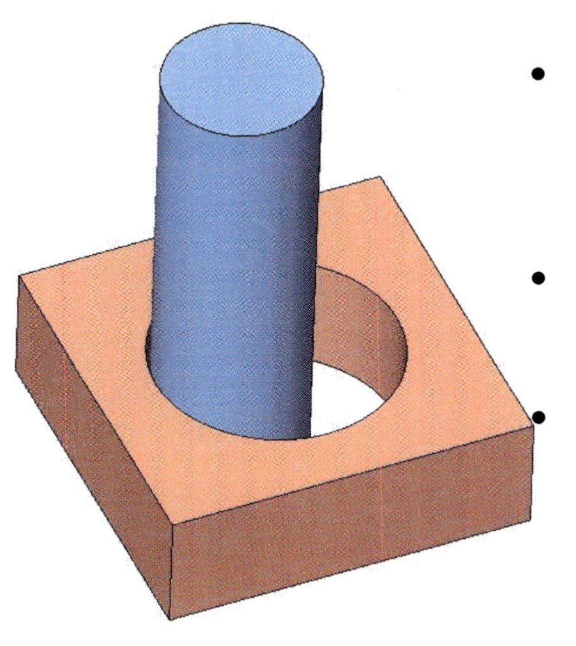

- Drilling is the fastest method of removing material. But when it's inconvenient or impractical to drill, holes can be circular milled.

- Holes can be milled from solid, or they can be pre-drilled.

- Helical Interpolation works best when milling to prevent a machined line down the side of the hole.
    - Helical motion involves XYZ simultaneous motion (*circular plus Z*)

Advanced CNC Mill Programming and Applied Mathematics Level 2

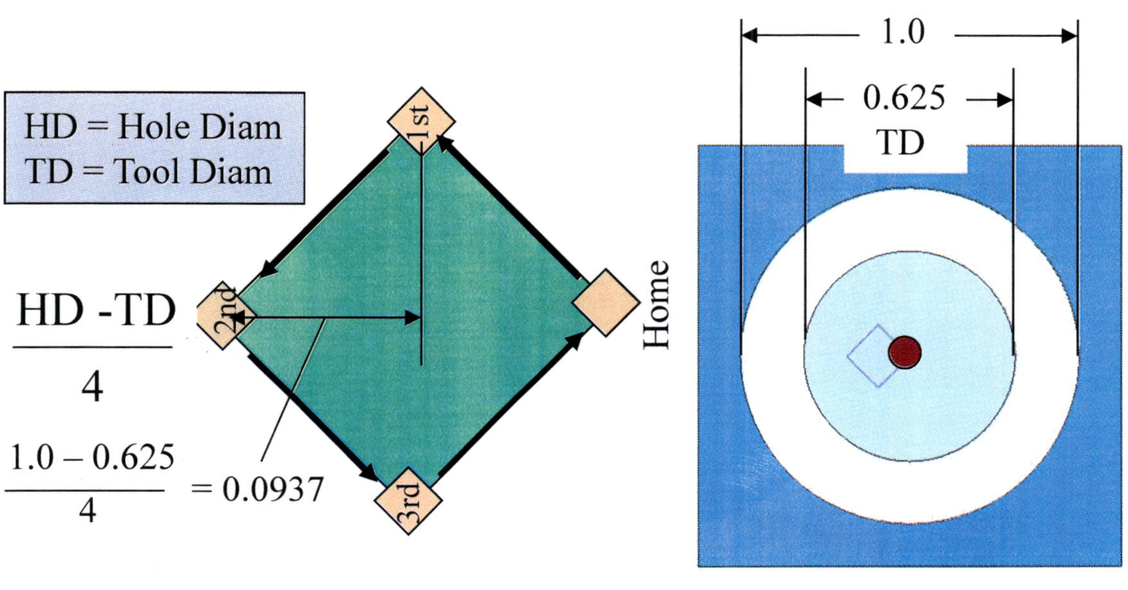

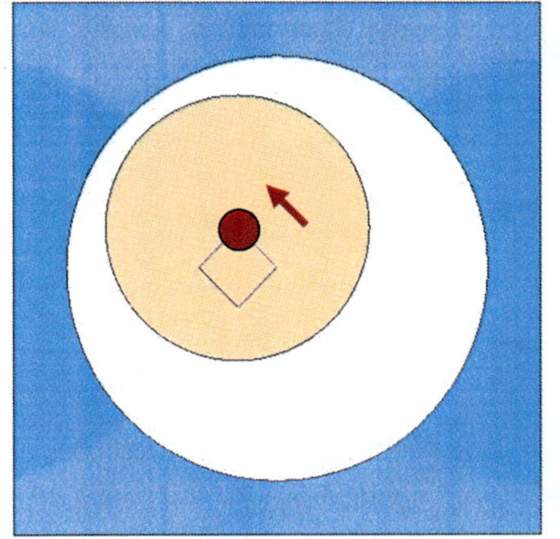

Step 1:

Linear Machine G1 to 1st Base.

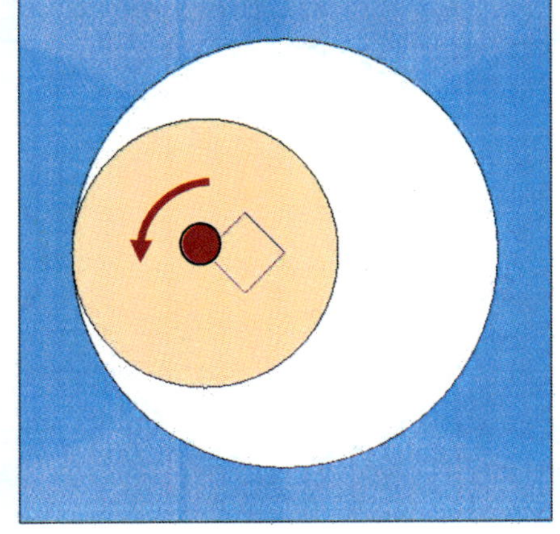

Step 2:

Circular Machine G3 to 2nd Base 45 degrees CCW.

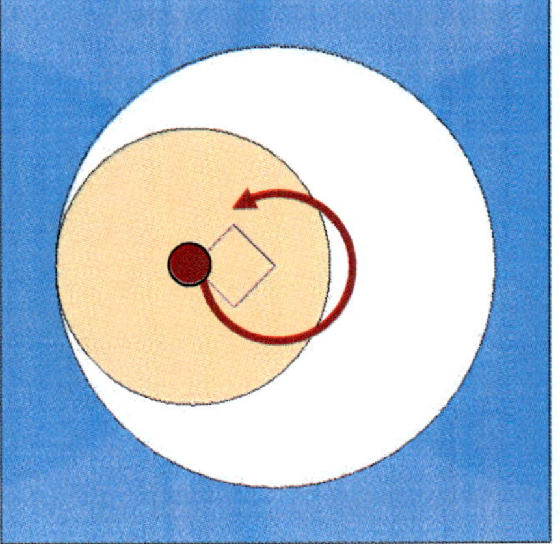

Step 3:

Circular Machine G3 360 degrees CCW.

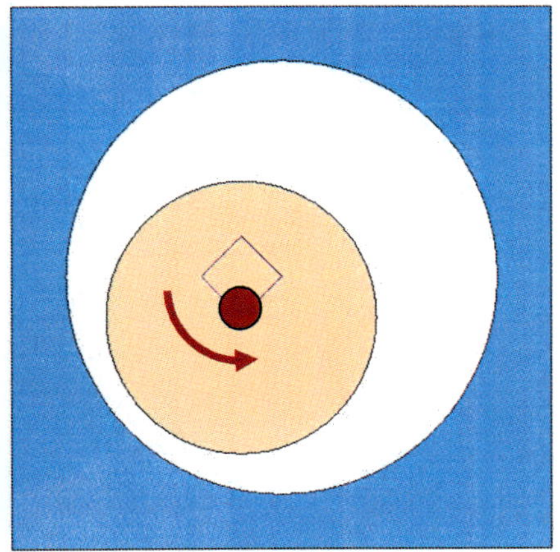

Step 4:

Circular Machine G3 to 3rd Base 45 degrees CCW. G1 to Home and repeat L'n' times.

INTERNAL CIRCULAR HOLE PROGRAM; TOP-DOWN; CLIMB

SPREAD SHEET PROGRAM

HOLE DIAMETER TO MACHINE:	1		Blue:	The user inputs data into the blue fields.
TOOL DIAMETER:	0.625		Rose:	The rose fields are automatically calculated.
DEPTH INCREMENT:	0.05		Tan:	The tan area represents the MAIN program.
HOLE DEPTH:	0.5		Green:	The green area represents the subroutine.
HOLE RADIUS = HD / 2	0.5		**INTERNAL CIRCULAR MILL LOGIC:**	
TOOL RADIUS = TD / 2	0.3125		*Prep*	Position the tool to the top-center of the hole to thread mill.
LONG LEG = HOME TO 2ND BASE	0.1875		*Prep*	Feed down programmed depth increment.
CNC G1 LEG = LONG LEG / 2	0.09375		Step 1:	Linear Interpolate to Location 1 (using calculated leg length)
SURFACE FEET:	400		Step 2:	Approach: G3 1/8 Turn to Loc 2 while dropping Depth Incr in Z
IPT:	0.005		Step 3:	G3 360 degrees with No Z
NUMBER OF FLUTES:	2		Step 4:	Exit: G3 1/8 Turn to Loc 4 No Z
RPM:	2445		Step 5:	G1 Back to Home Plate, then raise Z .100; Repeat as necessary
IPM:	24			

HOLE LOCATION:	X	0.000	Y	0.000
%				
O1010 (Circular Milling)				
T1 M6				
G54 G90 G0	X	0.000	Y	0.000
S	2445	M3		
G43 H1 Z.1				
M97	P100	L	10	
G28 G91 G0 Z0				
M30				

N100									(BEGIN SUBROUTINE	)
									(TOOL STARTS Z.100 ABOVE HOLE	)
G91									(INCREMENTAL MODE	)
G1		Z	-0.1000					F 24.4	(LOWER Z TO TOP OF HOLE	)
G1		X	-0.0938	Y	0.0938				(G1 TO 1ST BASE	)
G3		X	-0.0938	Y	-0.0938	Z	-0.0500	R 0.0938	(G3 ENTRY APPROACH	)
G3		I	0.1875	J	0				(FULL CIRCLE G3 CCW	)
G3		X	0.0938	Y	-0.0938	R	0.0938		(LEAD OUT G3 CCW	)
G1		X	0.0938	Y	0.0938				(G1 BACK TO HOME PLATE	)
G1		Z	0.100						(0.1 CLEAR FROM DEPTH OF CUT	)
G90									(RETURN TO ABSOLUTE MODE	)
M99									(RETURN TO PROGRAM	)
%										

Chapter 3
Corner Chamfering
Inside Diameter (ID)
Outside Diameter (OD)
Linear

Objectives

1. The student will list the benefits of chamfering with a chamfer tool.

2. The student will program inside and outside corner chamfers using   a chamfer tool and circular interpolation.

3. The student will calculate and program the final depth of the chamfer.

     Advanced CNC Mill Programming and Applied Mathematics Level 2

BENEFITS

The benefits of corner chamfering with a chamfer tool and a CNC machine include the following:

1. Chamfering allows you to put inside and outside beveled edges on parts.

2. CNC chamfering is much faster than conventional means.

3. CNC chamfering is much more accurate than conventional means.

4. CNC chamfering has the ability to better control the surface finish.

ID CORNER CHAMFERING

Sometimes it's necessary to **chamfer** a hole using circular interpolation and a chamfer tool. This is either called out on the part print, or simply added for a nice finishing touch. Chamfers can be any angle, but the most common angles are 30, 45, and 60 degrees.

Two calculations must be performed to program inside corner chamfers:

1) By applying the concepts from CNC Mill Level 1 (Drill Depths), the final Z depth is determined.
2) By applying the concepts from CNC Mill Level 2 (Circular Hole Milling), the math for the circular program is determined.

To avoid using the top corner of the tool, or the very bottom (center) of the tool, we define an effective tool diameter as somewhere in between. For example, when using a 0.500 diameter chamfer tool, we'll treat it as a 0.375 tool diameter to perform our calculations. This avoids using the outer most edge or the inner most point.

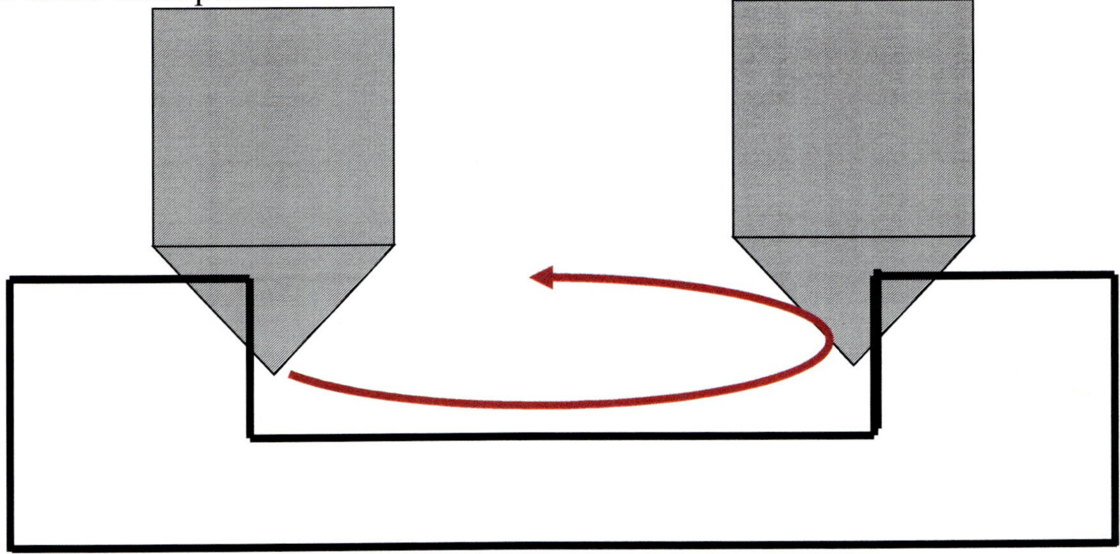

     Advanced CNC Mill Programming and Applied Mathematics Level 2

ID CORNER CHAMFERING

1. Z Depth is calculated by multiplying the *90 degree TPA (Tool Point Angle) factor* by the *effective tool diameter*. For example, 0.5 x 0.375 = 0.1875 deep.

2. To produce a 0.020 **inside corner chamfer** on the 3.0 diameter hole, we **_add_** 0.020 to each side of the circle, for a programmed diameter of **3.040**. Next we apply the concepts learned previously in *Circular Hole Milling* to calculate the leg size and IJ values.

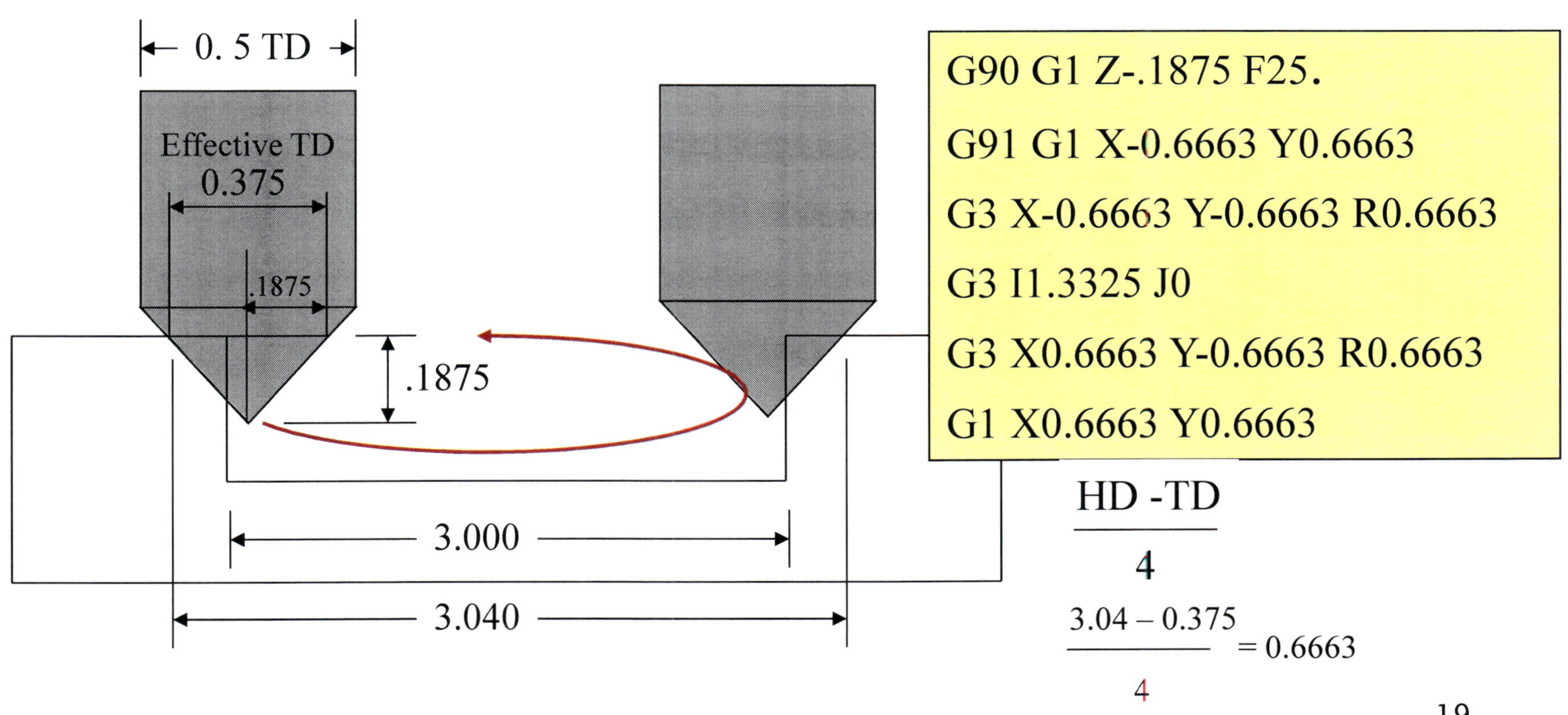

$$\frac{HD - TD}{4}$$

$$\frac{3.04 - 0.375}{4} = 0.6663$$

    Advanced CNC Mill Programming and Applied Mathematics Level 2

OD CORNER CHAMFERING

1. Z Depth is calculated in the same manner as inside corner chamfering—by multiplying the **90 degree TPA factor** by the ***effective tool diameter***. For example, 0.5 x 0.375 = 0.1875 deep.

2. To produce a 0.020 **outside corner chamfer** on the 3.0 diameter OD (e.g., coin), we ___subtract___ 0.020 from each side of the circle, for a programmed diameter of **2.960**. Next we apply the concepts learned previously in *Circular Hole Milling* to calculate the leg size and IJ values.

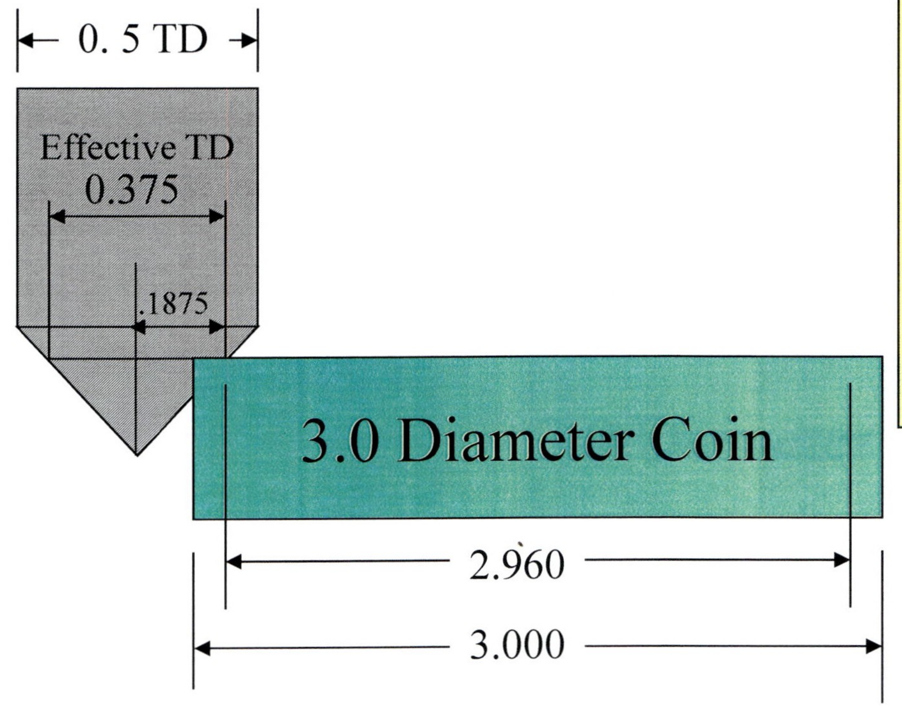

G90 G1 F25. Z-.1875

G91 G1 X-0.6463 Y0.6463

G3 X-0.6463 Y-0.6463 R0.6463

G3 I1.2925 J0

G3 X0.6463 Y-0.6463 R0.6463

G1 X0.6463 Y0.6463

$$\frac{HD - TD}{4}$$

$$\frac{2.96 - 0.375}{4} = 0.6463$$

     Advanced CNC Mill Programming and Applied Mathematics Level 2

LINEAR CORNER CHAMFERING

1. Z Depth is calculated in the same manner as ID and OD corner chamfering—by multiplying the **90 degree TPA factor** by the **effective tool diameter**. For example, 0.5 x 0.375 = 0.1875 deep.

2. To produce a 0.020 **outside corner chamfer** on a **4" square plate**, with ORIGIN defined at the center of the part, we **_subtract_** 0.020 from each side of the part, for a programmed edge of **1.98** instead of 2.0.

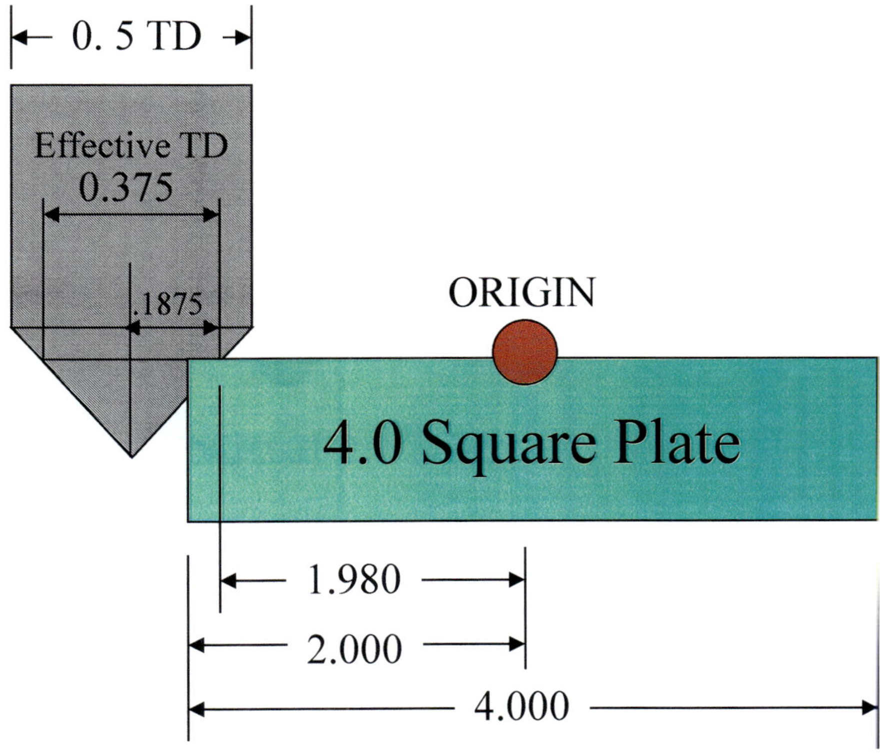

Magic Cube Project

Following is a CNC program that utilizes circular contouring to produce six holes per side, as well as six chamfers. All elements of this program have been taught in the previous pages.

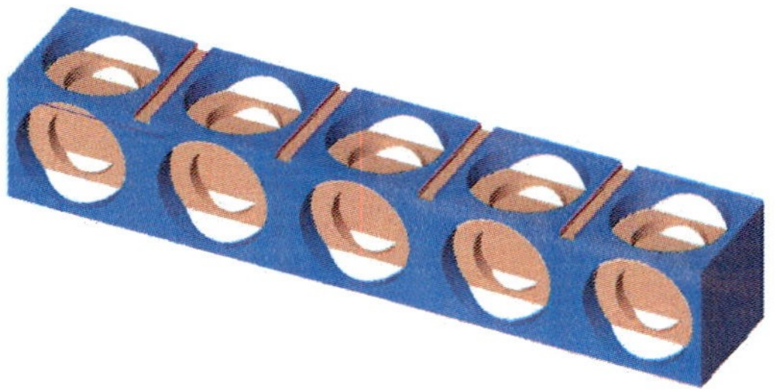

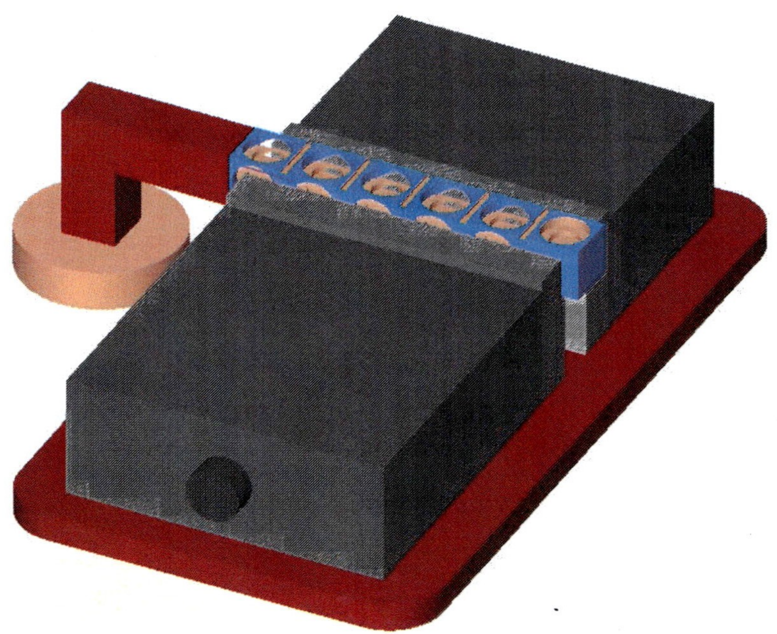

     Advanced CNC Mill Programming and Applied Mathematics Level 2

Main Program

```
%
O1060 (MAGIC CUBE  1 X 1 X 6.625)
G54 G90 G0 X0 Y0
( ######### HOLES ######### )
T3 M6        ( 3/8 END MILL )
S8000 M3
G90 G0 X0 Y0      ( CENTER OF HOLE )
G43 H3 G0 G90 Z.1 M8
M98 P1070 L5     ( FIVE HOLES )
( ######### CHAMFER ######### )
T16 M6  ( 1/2 CHAMFER TOOL )
S4600 M3
G90 G0 X0 Y0
G43 H16 G0 G90 Z.1 M8
M98 P1080 L5     ( FIVE CHAMFERS )
( ######## CUTOFF SLOTS ######### )
T8 M6  ( 1/8 END MILL )
S8000 M3
G90 G0 X1.125 Y-.6625 ( CENTER OF SLOT  )
G43 H8 G0 G90 Z.1 M8
M98 P1090 L4     ( FOUR SLOTS )
( ####### END OF PROGRAM  ########### )
G90 G0 X0 Y5.   ( OPERATOR LOAD POSITION )
G28 G91 G0 Z0
M30
%
```

Hole Subprogram

```
%
O1070 (MAGIC CUBE SUBPROGRAM 1070 HOLES )
G91 G1 Z-.098 F80.0 (BOTTOM HOLE FIRST)
G1 X-.0313 Y.0313
G3 X-.0313 Y-.0313 R.0313
M97 P100 L15
G3 I.0625 J0 (CLEANUP BOTTOM FACE)
G3 X.0313 Y-.0313 R.0313
G1 X.0313 Y.0313

G90 G0 Z0 (TOP HOLE NEXT)
G91 G1 X-.1093 Y.1093
G3 X-.1093 Y-.1093 R.1093
M97 P200 L6
G3 I.2185 J0 (CLEANUP BOTTOM FACE)
G3 X.1093 Y-.1093 R.1093
G1 X.1093 Y.1093
G90 G0 Z.1

G91 X1.125 (SHIFT TO NEXT HOLE)

N100 (HELICAL INTERP INSIDE HOLE)
G91 G3 I.0625 J0 Z-.02 (4 deg)
M99 (END SUBROUTINE 100)

N200 (HELICAL INTERP LARGER CIRCLE 0.812 DIA )
G91 G3 I.2185 J0 Z-.0325
M99 (end subroutine 200)
M99 (END SUBPRO 1070 HOLES )
%
```

Main Program

```
%
O1060 (MAGIC CUBE  1 X 1 X 6.625)
G54 G90 G0 X0 Y0
( ######### HOLES ######### )
T3 M6        ( 3/8 END MILL )
S8000 M3
G90 G0 X0 Y0      ( CENTER OF HOLE )
G43 H3 G0 G90 Z.1 M8
M98 P1070 L6      ( FIVE HOLES )
( ######### CHAMFER ######### )
T16 M6  ( 1/2 CHAMFER TOOL )
S4600 M3
G90 G0 X0 Y0
G43 H16 G0 G90 Z.1 M8
M98 P1080 L6      ( FIVE CHAMFERS )
( ######## CUTOFF SLOTS ######### )
T8 M6  ( 1/8 END MILL )
S8000 M3
G90 G0 X1.125 Y-.6625 ( CENTER OF SLOT  )
G43 H8 G0 G90 Z.1 M8
M98 P1090 L4      ( FOUR SLOTS )
( ####### END OF PROGRAM  ########### )
G90 G0 X0 Y5.    (OPERATOR LOAD POSITION )
G28 G91 G0 Z0
M30
%
```

% Chamfer Subprogram

```
O1080 (MAGIC CUBE SUBPRO 1080 CHAMFER )
( ## 0.005 CHAMFER  ## )
( BOTTOM HOLE 1ST )
(TREAT AS .2 TD; DP = 0.1875 + .1 )
( TREAT AS 0.500 DIAMETER; 0.005 PER SIDE )

G90 G1 Z-.2875 F25.
G91 G1 X-.075 Y.075
G3 X-.075 Y-.075 R.075
G3 I.15 J0
G3 X.075 Y-.075 R.075
G1 X.075 Y.075
( TOP HOLE 2ND )
(TREAT AS .375 TD; DP = 1/2 TD )
( TREAT AS .812 DIAMETER; 0.005 PER SIDE)
G90 Z-.1875  (1/2 TD )
G91 X-.1093 Y.1093
G3 X-.1093 Y-.1093 R.1093
G3 I.2186 J0
G3 X.1093 Y-.1093 R.1093
G1 X.1093 Y.1093
G90 Z.1

G91 X1.125 (SHIFT TO NEXT HOLE)
M99 (END SUBPRO 1080 CHAMFER )
%
```

Main Program

%

O1060 (MAGIC CUBE 1 X 1 X 6.625)

G54 G90 G0 X0 Y0

(######### HOLES #########)

T3 M6   (3/8 END MILL)

S8000 M3

G90 G0 X0 Y0   (CENTER OF HOLE)

G43 H3 G0 G90 Z.1 M8

M98 P1070 L6   (FIVE HOLES)

(######### CHAMFER #########)

T16 M6 (1/2 CHAMFER TOOL)

S4600 M3

G90 G0 X0 Y0

G43 H16 G0 G90 Z.1 M8

M98 P1080 L6   (FIVE CHAMFERS)

(######## CUTOFF SLOTS #########)

T8 M6 (1/8 END MILL)

S8000 M3

G90 G0 X1.125 Y-.6625 (CENTER OF SLOT)

G43 H8 G0 G90 Z.1 M8

M98 P1090 L4   (FOUR SLOTS)

(####### END OF PROGRAM   ###########)

G90 G0 X0 Y5.   (OPERATOR LOAD POSITION)

G28 G91 G0 Z0

M30

%

% Slots Sub Program

O1090 (MAGIC CUBE SUB PRO 1090 SLOTS)

(## SLOTS ##)

G90 G1 Z-.033 F20.

G1 Y.6625

G0 Z.1

G0 Y-.6625

G1 Z-.066

G1 Y.6625

G0 Z.1

G0 Y-.6625

G1 Z-.1

G1 Y.6625

G0 Z.1

G91 X1.125 (SHIFT TO NEXT HOLE)

M99 (END SUB PRO 1090 SLOTS)

%

Chapter 4
Thread Milling

External Right-hand
External Left-hand
Internal Right-hand
Internal Left-hand

    Advanced CNC Mill Programming and Applied Mathematics Level 2

Objectives

1. The student will list the benefits of thread milling.

2. The student will demonstrate knowledge of thread terminology.

3. The student will develop thread milling programs for internal and external threads.

Thread Milling

- **Thread Milling is one of the fastest, most effective and cost-efficient method for producing threads.**

- **Requires Helical Interpolation**
  - XYZ simultaneous motion (*circular plus Z*)

- **Capable of producing:**
  - Internal, External, Right-hand, and Left-hand threads of all sizes—especially nonstandard diameters.

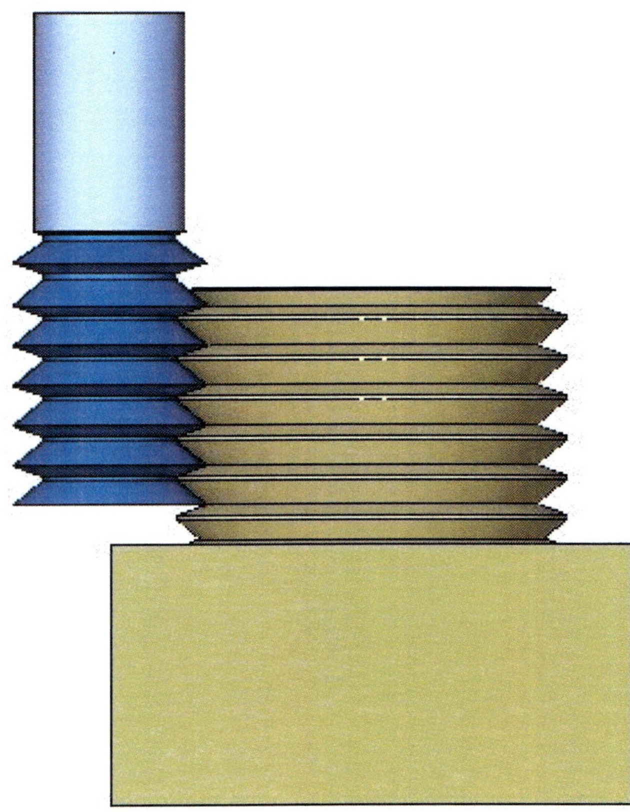

    Advanced CNC Mill Programming and Applied Mathematics Level 2

Thread Milling Benefits

- Same tool can be used for external and internal threads

- Same tool can be used for right-hand and left-hand threads

- 100% thread depth can be achieved

- Threat of broken tap eliminated

- 75% less HP required than tapping

- One tool fits multiple diameters

    Advanced CNC Mill Programming and Applied Mathematics Level 2

Thread Mill Considerations

- Internal or External thread

- Right-hand or Left-hand

- Major and Minor Diameters

- Threads Per Inch (TPI)

- Tap Drill Size (internal threads only)

- May want to make multiple passes on difficult to machine materials

- Arc-in and Arc-out. Include Z. For example, a 90 degree arc would need ¼ Pitch in Z.

- RPM and IPM are calculated following the Drill and Mill formulae.

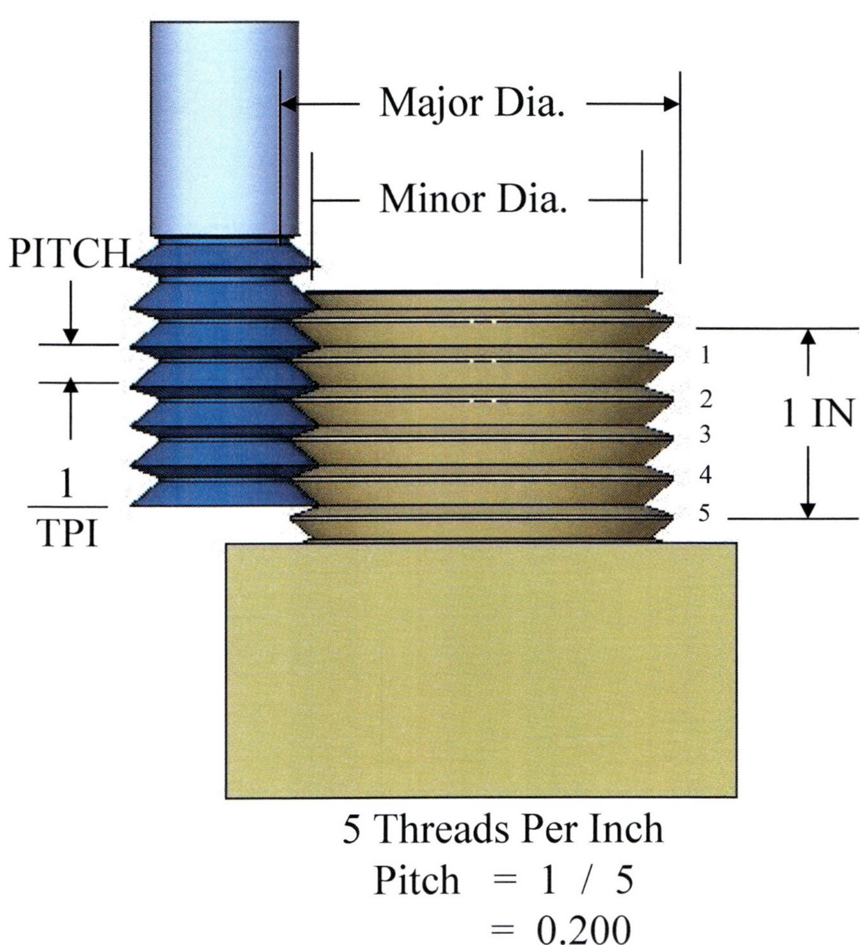

5 Threads Per Inch
Pitch = 1 / 5
       = 0.200

   Advanced CNC Mill Programming and Applied Mathematics Level 2

Thread Milling (CLIMB)

EXTERNAL

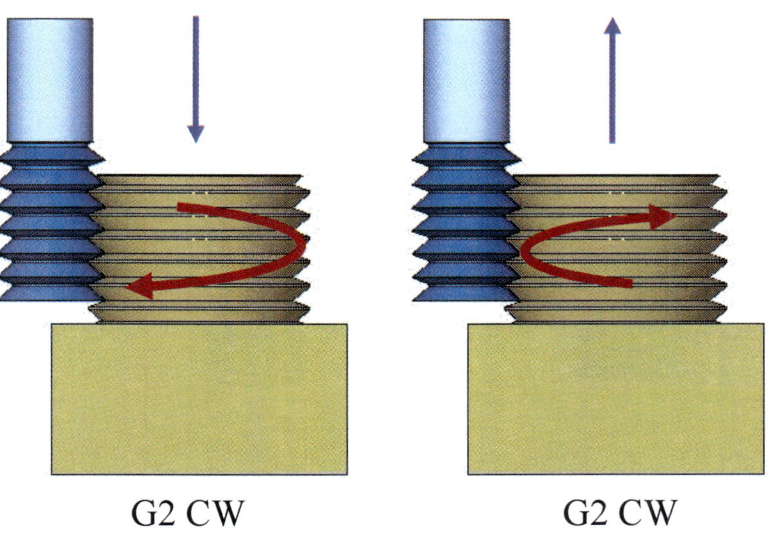

INTERNAL

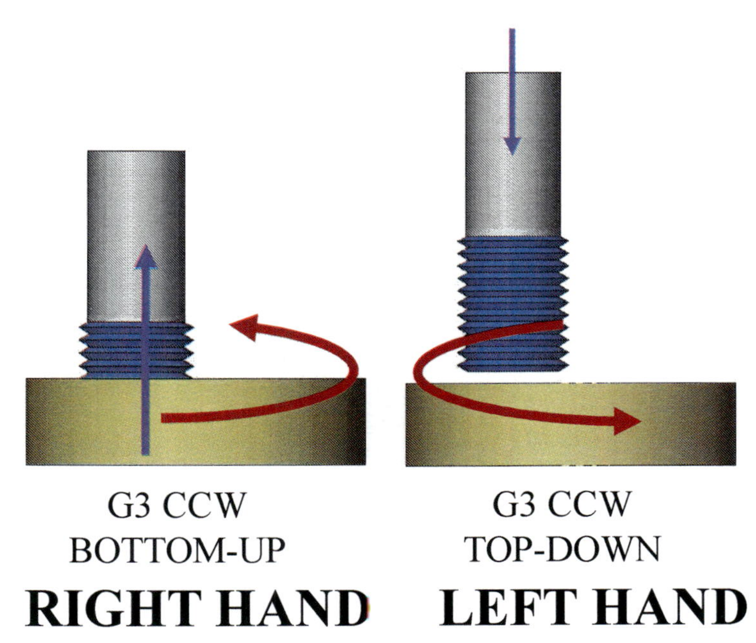

     Advanced CNC Mill Programming and Applied Mathematics Level 2

Internal RH Thread Milling

First, drill or circular mill the hole prior to internal thread milling. The size of the hole is referred to as **"Tap Drill Size"**. The tap drill size can be found in Tap Drill charts up to 1.0 inches. Anything larger must be calculated as shown below.

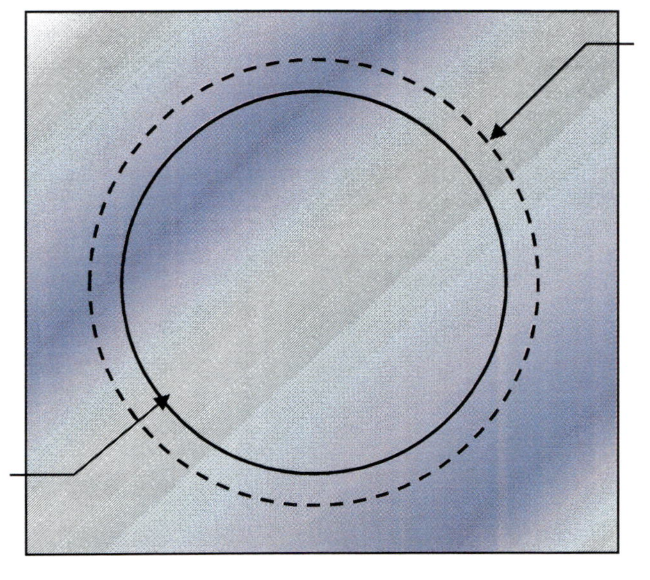

$$\text{Tap Drill Size} = \text{Thread Diameter} - \frac{0.01299 \times \%\ of\ Full\ Thread}{\text{Number of Threads Per Inch}}$$

Internal RH Thread Milling - Climb

- Position the thread mill to the *bottom-center* of the hole.

- Calculate leg length to be used for pre-entry.

  Leg Length = (Major Diam – Tool Diameter) / 4

- Linear Contour (G1) to location 1.
  X-(*leg length*) Y+(*leg length*)

- Arc On 45° CCW while raising Z 1/8 Pitch.
  (Pitch = 1 / TPI)

- Circular Interpolate 360 CCW while raising Z one Pitch.
  Note, this is assuming you have a thread mill capable of completing the thread in one revolution. Otherwise you'll have to perform another rotation.

- Arc Off 45° CCW while raising Z 1/8 Pitch).

- Retract to clearance plane.

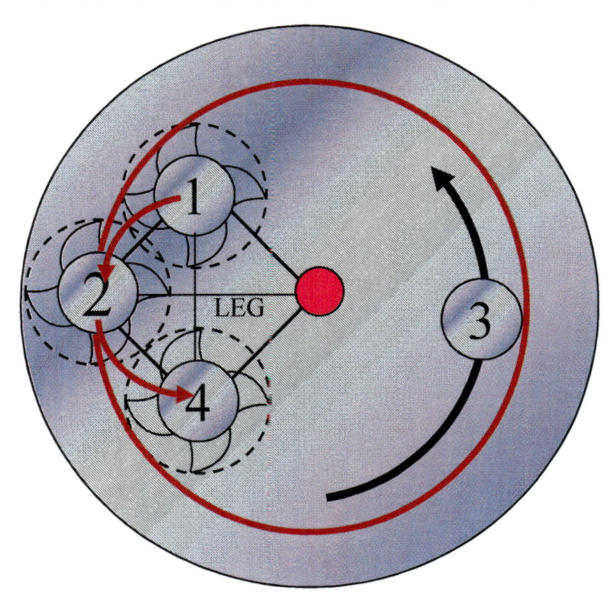

Advanced CNC Mill Programming and Applied Mathematics Level 2

INTERNAL RIGHTHAND THREAD MILL PROGRAM; Bottom-Up CCW - CLIMB

SPREAD SHEET PROGRAM

THREAD SIZE:	1		Blue:	The user inputs data into the blue fields.	
THREADS PER INCH (TPI):	10		Rose:	The rose fileds are automatically calculated.	
EFFECTIVE TOOL DIA:	0.5		Tan:	The tan area represents the MAIN program.	
HOLE DEPTH:	0.5		Green:	The green area represents the subroutine.	
PITCH: (1/TPI)	0.1		INTERNAL THREAD MILL LOGIC:		
CNC G1 LEG:	0.125		Prep	Position the tool to the bottom-center of the hole to thread mill.	
SURFACE FEET:	400		Prep	Feed down to bottom of hole	
IPT:	0.0015		Step 1:	Linear Interpolate to Location 1 (using calculated leg length)	
NUMBER OF FLUTES:	3		Step 2:	Approach: G3 1/8 Turn to Loc 2 while raising 1/8 Pitch in Z	
RPM:	3056		Step 3:	G3 360 degrees while raising 1 Pitch in Z	
IPM:	14		Step 4:	Exit: G3 1/8 Turn to Loc 4 while raising 1/8 Pitch in Z	
Tap Drill Size (65% Thread):	0.9156	(FYI Only)	Return to center of hole.		
MILLING LOCATION:	0	0			

%				
O1010 (THREAD MILL)				
T1 M6				
G54 G90 G0	X	0	Y	0
S	3056	M3		
G43 H1	G0	Z.1		
M97	P100			
G28 G91 G0 Z0				
M30				

N100									(BEGIN SUBROUTINE	)
									(START AT TOP CENTER OF HOLE	)
G91									(INCREMENTAL MODE	)
G1		Z	-0.5					F 13.8	(LOWER Z TO BOTTOM OF HOLE	)
G1		X	-0.125	Y	0.125				(POSITION TO TOP OF TRIANGLE	)
G3		X	-0.125	Y	-0.125	R	0.125	Z 0.0125	(RAMP ON - RAISE Z 1/8 PITCH	)
G3		I	0.25	J	0	Z	0.1		(FULL CIRCLE - RAISE Z 1 PITCH	)
G3		X	0.125	Y	-0.125	R	0.125	Z 0.0125	(RAMP OFF - RAISE Z 1/8 PITCH	)
G1		X	0.125	Y	0.125				(RETURN TO START LOCATION	)
G1		Z	0.5						(CLEARANCE	)
G90									(RETURN TO ABSOLUTE MODE	)
M99									(RETURN TO PROGRAM	)
%										

NOTE ON APPLYING CUTTER DIAMETER COMPENSATION:

The calculated leg =	0.125	
The hypotenuse =	0.176777	
1/2 the Tool Diameter =	0.25	***This represents the minimum length of travel for applying CDC***
		- if the hypotenuse is great than or = to this, G41 is applied in line 32 and cancelled in line 36.
		- else you cannot apply CDC

Sample Internal RH Thread Milling

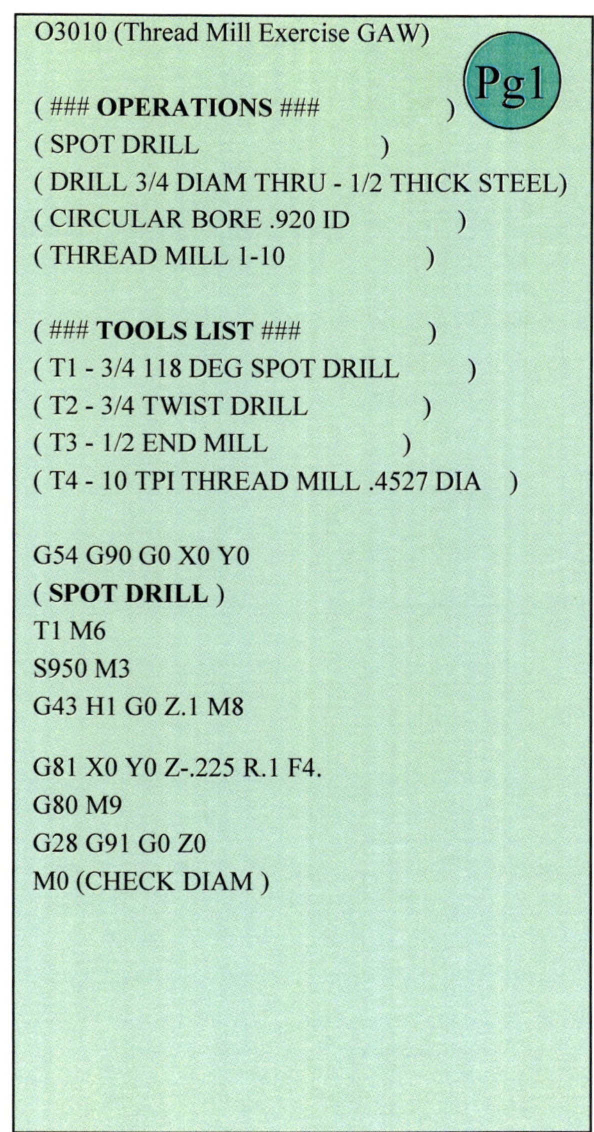

```
O3010 (Thread Mill Exercise GAW)

( ### OPERATIONS ###          )
( SPOT DRILL                  )
( DRILL 3/4 DIAM THRU - 1/2 THICK STEEL)
( CIRCULAR BORE .920 ID       )
( THREAD MILL 1-10            )

( ### TOOLS LIST ###          )
( T1 - 3/4 118 DEG SPOT DRILL     )
( T2 - 3/4 TWIST DRILL            )
( T3 - 1/2 END MILL               )
( T4 - 10 TPI THREAD MILL .4527 DIA   )


G54 G90 G0 X0 Y0
( SPOT DRILL )
T1 M6
S950 M3
G43 H1 G0 Z.1 M8

G81 X0 Y0 Z-.225 R.1 F4.
G80 M9
G28 G91 G0 Z0
M0 (CHECK DIAM )
```

Pg1

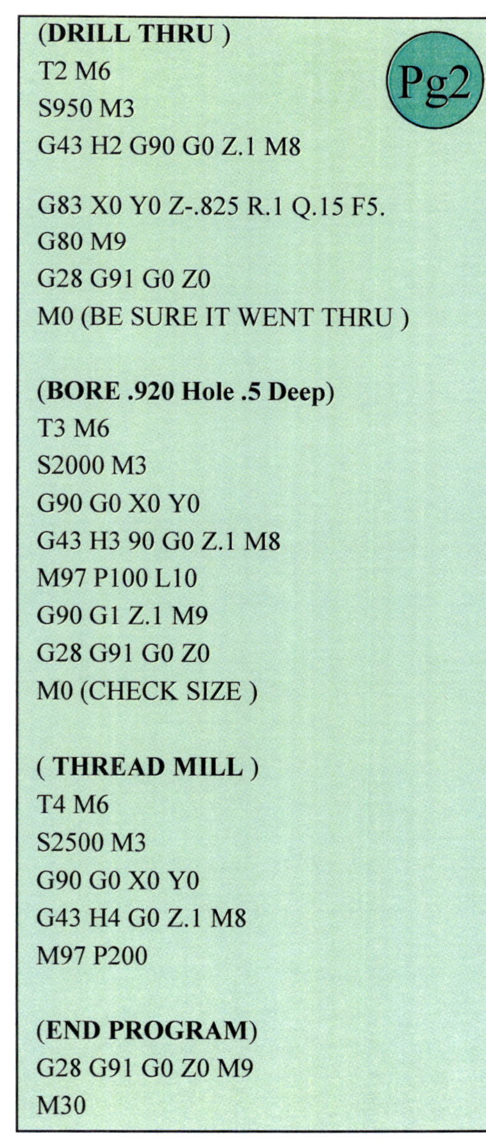

```
(DRILL THRU )
T2 M6
S950 M3
G43 H2 G90 G0 Z.1 M8

G83 X0 Y0 Z-.825 R.1 Q.15 F5.
G80 M9
G28 G91 G0 Z0
M0 (BE SURE IT WENT THRU )

(BORE .920 Hole .5 Deep)
T3 M6
S2000 M3
G90 G0 X0 Y0
G43 H3 90 G0 Z.1 M8
M97 P100 L10
G90 G1 Z.1 M9
G28 G91 G0 Z0
M0 (CHECK SIZE )

( THREAD MILL )
T4 M6
S2500 M3
G90 G0 X0 Y0
G43 H4 G0 Z.1 M8
M97 P200

(END PROGRAM)
G28 G91 G0 Z0 M9
M30
```

Pg2

Subroutines:

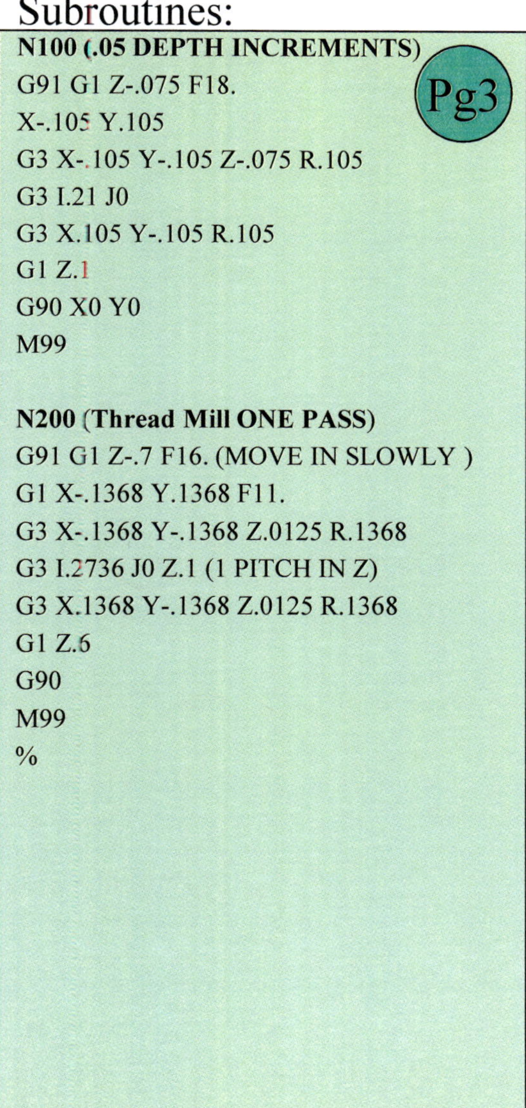

```
N100 (.05 DEPTH INCREMENTS)
G91 G1 Z-.075 F18.
X-.105 Y.105
G3 X-.105 Y-.105 Z-.075 R.105
G3 I.21 J0
G3 X.105 Y-.105 R.105
G1 Z.1
G90 X0 Y0
M99


N200 (Thread Mill ONE PASS)
G91 G1 Z-.7 F16. (MOVE IN SLOWLY )
G1 X-.1368 Y.1368 F11.
G3 X-.1368 Y-.1368 Z.0125 R.1368
G3 I.2736 J0 Z.1 (1 PITCH IN Z)
G3 X.1368 Y-.1368 Z.0125 R.1368
G1 Z.6
G90
M99
%
```

Pg3

  Advanced CNC Mill Programming and Applied Mathematics Level 2

Internal LH Thread Milling - Climb

Thread mill **counterclockwise** G3 following the **Major Thread Diameter**. Machine in a **top-down** direction using a helical motion. Use Cutter Compensation (CC) Left (G41) and store a zero tool diameter in the offset page, allowing the operator to *comp* to fit.

- The steps for milling a left-hand internal thread are similar to RH except that the thread direction is top-down instead of bottom-up.
  - This maintains Climb Cutting
  - You could instead cut bottom-up using G2 using conventional machining.
- Pay special attention to lower the Z inside the hole to a point 1-1/8 pitch above the bottom if machining only one revolution to insure a complete thread.

    Advanced CNC Mill Programming and Applied Mathematics Level 2

<table>
<tr><td colspan="11">INTERNAL LEFTHAND THREAD MILL PROGRAM; Top-Down CCW - CLIMB</td></tr>
<tr><td colspan="11">SPREAD SHEET PROGRAM</td></tr>
</table>

THREAD SIZE:	1		Blue:	The user inputs data into the blue fields.
THREADS PER INCH (TPI):	10		Rose:	The rose fileds are automatically calculated.
EFFECTIVE TOOL DIA:	0.5		Tan:	The tan area represents the MAIN program.
HOLE DEPTH:	0.5		Green:	The green area represents the subroutine.
PITCH: (1/TPI)	0.1			**INTERNAL THREAD MILL LOGIC:**
CNC G1 LEG:	0.125		*Prep*	Position the tool to the top-center of the hole
SURFACE FEET:	400		NOTE:	*Lower Z 1-1/8 Pitch from bottom*
IPT:	0.0015		Step 1:	Linear Interpolate to Location 1 (using calculated leg length)
NUMBER OF FLUTES:	3		Step 2:	Approach: G3 1/8 Turn to Loc 2 while dropping 1/8 Pitch in Z
RPM:	3056		Step 3:	G3 360 degrees while dropping 1 Pitch in Z
IPM:	14		Step 4:	Exit: G3 1/8 Turn to Loc 4 while dropping 1/8 Pitch in Z
Tap Drill Size (65% Thread):	0.9156 *(FYI Only)*			Return to center of hole.
MILLING LOCATION:	0	0		

%				
O1010 (THREAD MILL)				
T1 M6				
G54 G90 G0	X	0	Y	0
S	3056	M3		
G43 H1	G0	Z.1		
M97	P100			
G28 G91 G0 Z0				
M30				

N100								(BEGIN SUBROUTINE	)
								(START AT TOP CENTER OF HOLE	)
G91								(INCREMENTAL MODE	)
G1		Z -0.4				F 13.8		(LOWER Z TO BOTTOM OF HOLE	)
G1		X -0.125	Y 0.125					(POSITION TO TOP OF TRIANGLE	)
G3		X -0.125	Y -0.125	R 0.125	Z -0.0125			(RAMP ON - RAISE Z 1/8 PITCH	)
G3		I 0.25	J 0	Z -0.1				(FULL CIRCLE - RAISE Z 1 PITCH	)
G3		X 0.125	Y -0.125	R 0.125	Z -0.0125			(RAMP OFF - RAISE Z 1/8 PITCH	)
G1		X 0.125	Y 0.125					(RETURN TO START LOCATION	)
G1		Z 0.5						(CLEARANCE	)
G90								(RETURN TO ABSOLUTE MODE	)
M99								(RETURN TO PROGRAM	)
%									

NOTE ON APPLYING CUTTER DIAMETER COMPENSATION:

The calculated leg =	0.125	
The hypotenuse =	0.176777	
1/2 the Tool Diameter =	0.25	***This represents the minimum length of travel for applying CDC***
		- if the hypotenuse is great than or = to this, G41 is applied in line 32 and cancelled in line 36.
		- else you cannot apply CDC

    Advanced CNC Mill Programming and Applied Mathematics Level 2

External RH Thread Milling

$$\text{Thread Depth} = \frac{0.6495}{\text{Number of Threads}}$$

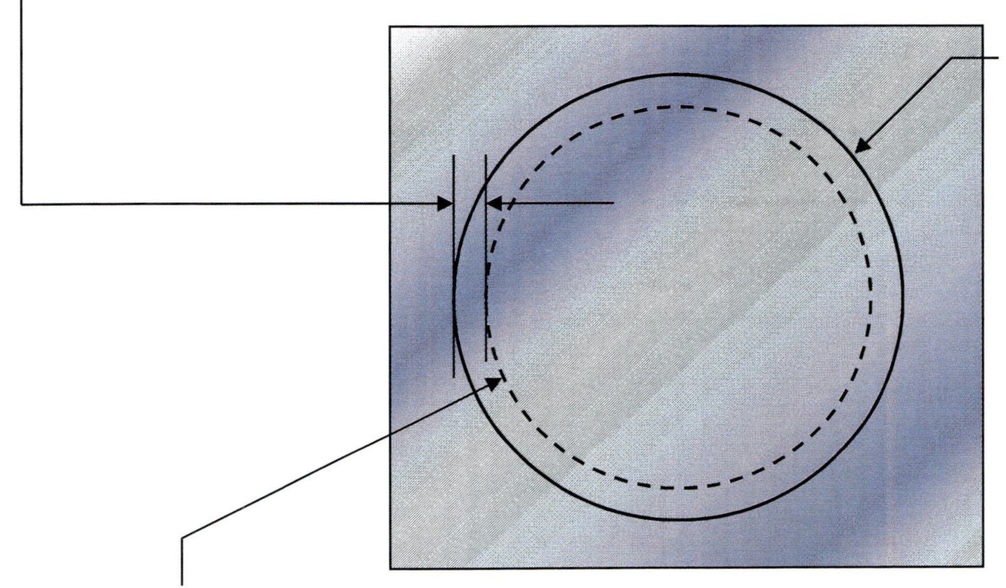

Major Diameter is machined to size by contouring G2 (climb) with an end mill.   For example: **2.25 – 10:**   Major Diameter = 2.25

When thread milling external threads, you'll contour the Minor Diameter.
Minor Diameter = Major Diameter – (2 x Thread Depth)

    Advanced CNC Mill Programming and Applied Mathematics Level 2

External RH Thread Milling - Climb

Thread Depth:

$$0.06495 = \frac{0.6495}{10}$$

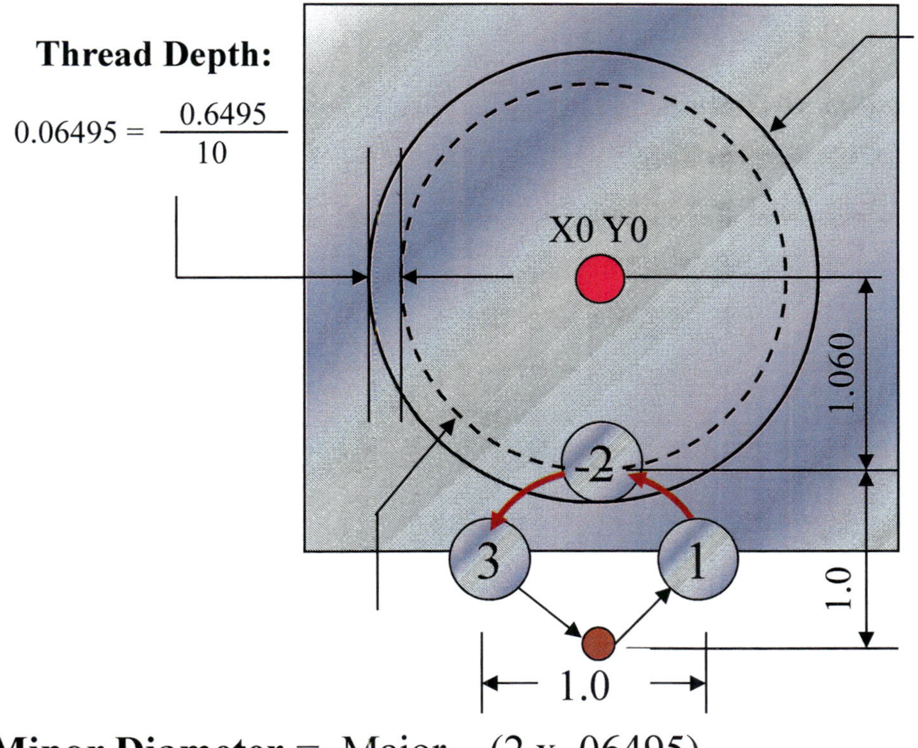

Minor Diameter = Major − (2 x .06495)
= 2.120

2.25 - 10

- Position to Home Plate and ¼ Pitch above top.
- Linear machine to 1st base (45°). No Z.
- G3 90° to 2nd base; drop Z ¼ Pitch. while
- G2 360° Minor Diameter; drop Z one Pitch.
- G3 90° to 3rd base; drop Z ¼ Pitch.
- Linear machine to Home (45°).

G0 G90 X0 Y-2.060	
G1 G41 D1 X0.5 Y-1.560	(1st Base)
G3 X0 Y-1.060 R.5 Z-0.025	(2nd Base)
G2 I0 J1.060 Z-0.1	(360 degrees CW)
G3 X-0.5 Y-1.560 R.5 Z-0.025	(3rd Base)
G1 G40 X0 Y-2.060	(Home Plate)

     Advanced CNC Mill Programming and Applied Mathematics Level 2

EXTERNAL RIGHTHAND THREAD MILL PROGRAM; Top-Down - CW - CLIMB
SPREAD SHEET PROGRAM

MAJOR THREAD DIAM:	2.25	Blue:	The user inputs data into the blue fields.
THREADS PER INCH (TPI):	10	Rose:	The rose fileds are automatically calculated.
Thread Depth (.6495/Thrds)	0.06495	Tan:	The tan area represents the MAIN program.
MINOR THREAD DIAM:	2.1201	Green:	The green area represents the subroutine.
NOMINAL BASEBALL DIAM	1		**INTERNAL THREAD MILL LOGIC:**
HOLE DEPTH:	0.5		Position the tool to Home Plate; Z 1/4 Pitch above part
PITCH: (1/TPI)	0.1		Feed to 1st Base G1; Turn on CDC Left G41
CNC G1 LEG:	0.5		G3 to 2nd base; drop Z 1/4 Pitch
SURFACE FEET:	400		G2 360; drop Z 1 Pitch
IPT:	0.0015		G3 90; drop Z 1/4 Pitch
NUMBER OF FLUTES:	3		Feed to Home Plate G1, Turn off CDC G40
RPM:	1528		
IPM:	7		
MILLING LOCATION:	0	0	

%					
O1030 (THREAD MILL)					
T1 M6					
G54 G90 G0	X	0	Y	-2.0601	
S	1528	M3			
G43 H1	G0	Z	0.025		
M97	P100				
G28 G91 G0 Z0					
M30					

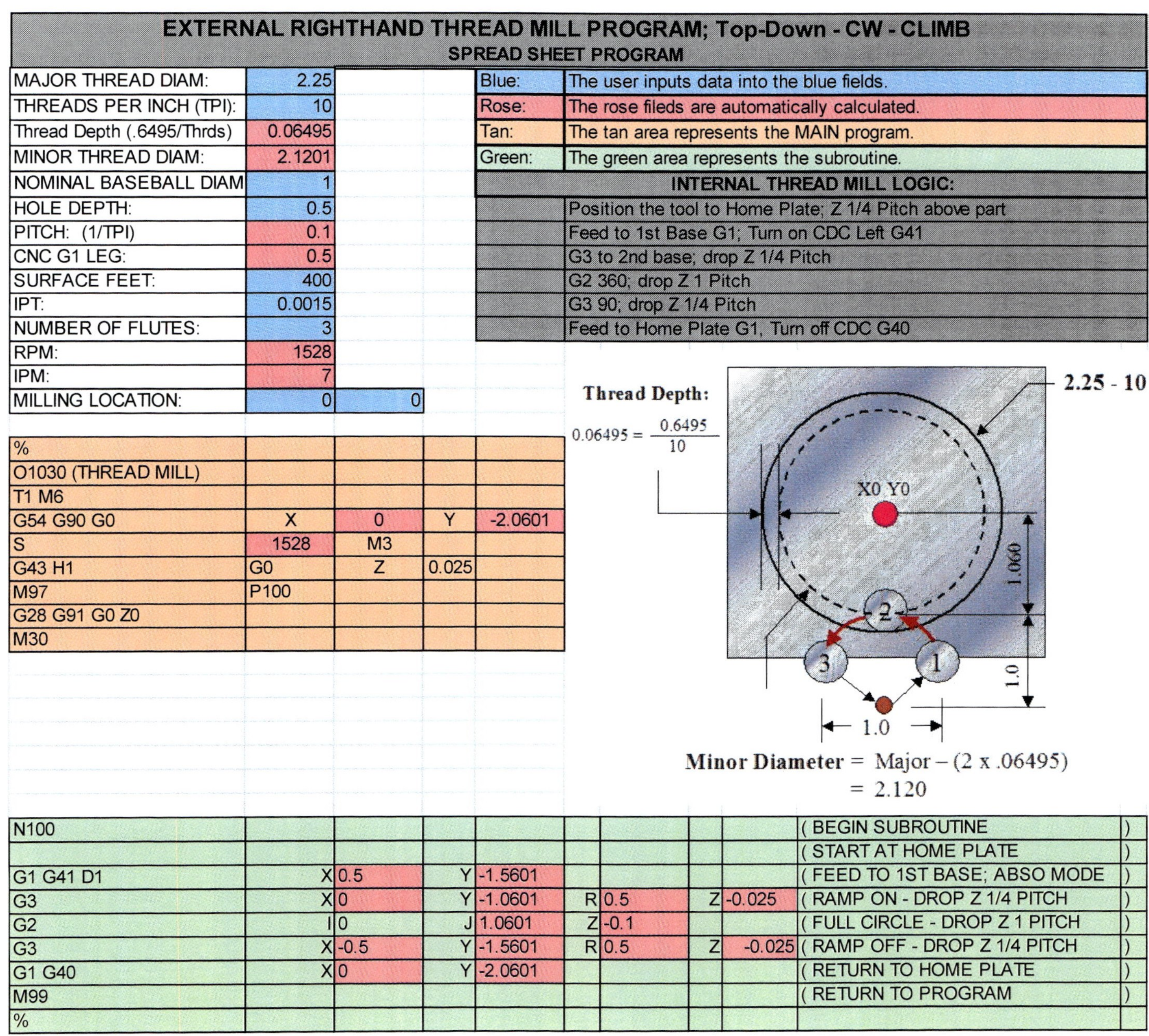

N100								(BEGIN SUBROUTINE	)
								(START AT HOME PLATE	)
G1 G41 D1	X	0.5	Y	-1.5601				(FEED TO 1ST BASE; ABSO MODE	)
G3	X	0	Y	-1.0601	R	0.5	Z -0.025	(RAMP ON - DROP Z 1/4 PITCH	)
G2	I	0	J	1.0601	Z	-0.1		(FULL CIRCLE - DROP Z 1 PITCH	)
G3	X	-0.5	Y	-1.5601	R	0.5	Z -0.025	(RAMP OFF - DROP Z 1/4 PITCH	)
G1 G40	X	0	Y	-2.0601				(RETURN TO HOME PLATE	)
M99								(RETURN TO PROGRAM	)
%									

     Advanced CNC Mill Programming and Applied Mathematics Level 2

External LH Thread Milling - Climb

Thread mill **clockwise** G2 following the **Major Thread Diameter**. Machine in a **bottom-up** direction using a helical motion. Use Cutter Compensation (CC) Left (G41) and store a zero tool diameter in the offset page, allowing the operator to *comp* to fit.

- The steps for milling a left-hand external thread are similar to RH except that the thread is bottom-up instead of top-down.
  - This maintains Climb Cutting
  - You could instead cut top-down using G3 using conventional machining.

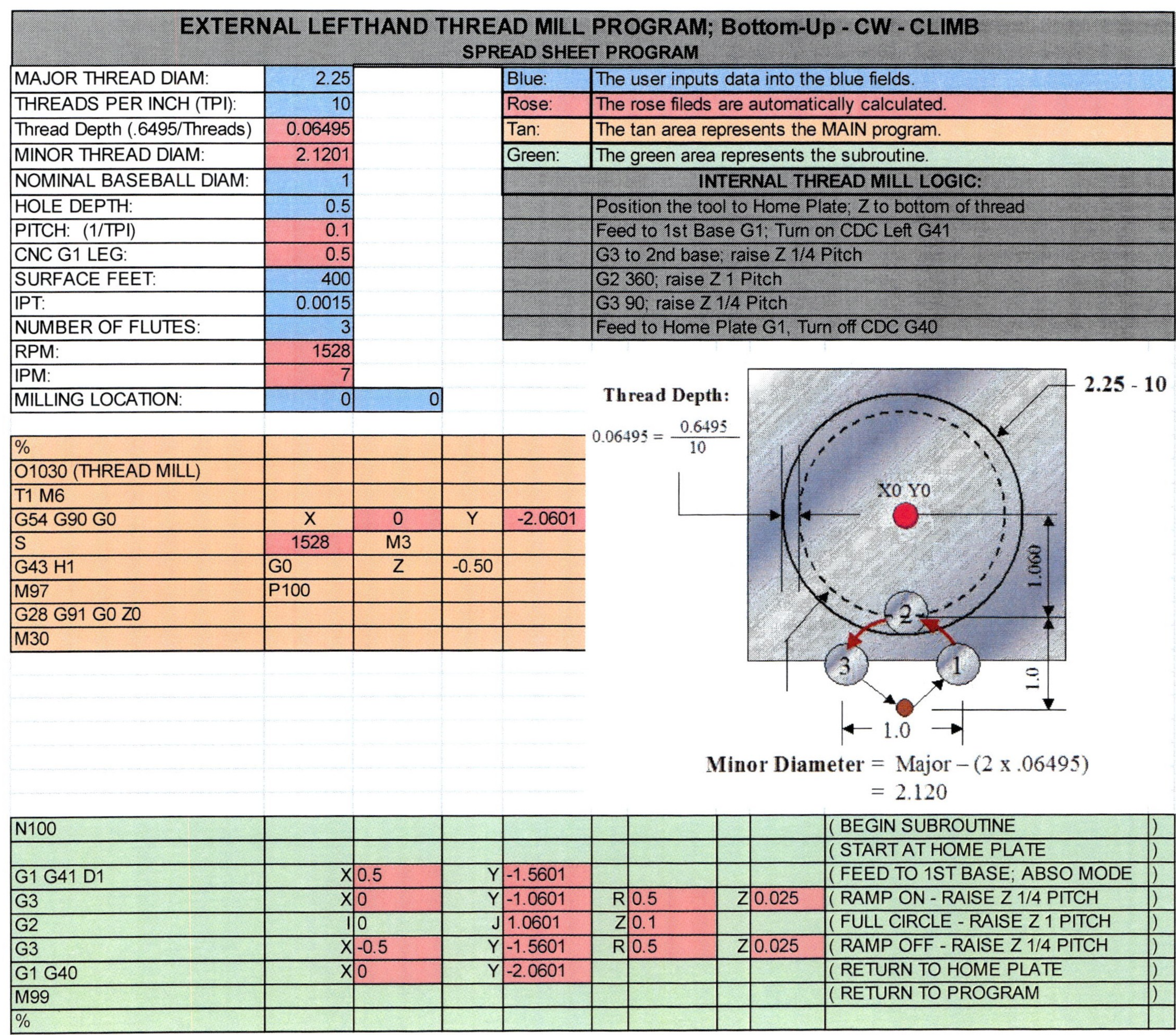

EXTERNAL LEFTHAND THREAD MILL PROGRAM; Bottom-Up - CW - CLIMB
SPREAD SHEET PROGRAM

MAJOR THREAD DIAM:	2.25		Blue:	The user inputs data into the blue fields.	
THREADS PER INCH (TPI):	10		Rose:	The rose fileds are automatically calculated.	
Thread Depth (.6495/Threads)	0.06495		Tan:	The tan area represents the MAIN program.	
MINOR THREAD DIAM:	2.1201		Green:	The green area represents the subroutine.	
NOMINAL BASEBALL DIAM:	1			**INTERNAL THREAD MILL LOGIC:**	
HOLE DEPTH:	0.5			Position the tool to Home Plate; Z to bottom of thread	
PITCH: (1/TPI)	0.1			Feed to 1st Base G1; Turn on CDC Left G41	
CNC G1 LEG:	0.5			G3 to 2nd base; raise Z 1/4 Pitch	
SURFACE FEET:	400			G2 360; raise Z 1 Pitch	
IPT:	0.0015			G3 90; raise Z 1/4 Pitch	
NUMBER OF FLUTES:	3			Feed to Home Plate G1, Turn off CDC G40	
RPM:	1528				
IPM:	7				
MILLING LOCATION:	0	0			

%				
O1030 (THREAD MILL)				
T1 M6				
G54 G90 G0	X	0	Y	-2.0601
S	1528	M3		
G43 H1	G0	Z	-0.50	
M97	P100			
G28 G91 G0 Z0				
M30				

N100							(BEGIN SUBROUTINE	)
							(START AT HOME PLATE	)
G1 G41 D1		X 0.5	Y -1.5601				(FEED TO 1ST BASE; ABSO MODE	)
G3		X 0	Y -1.0601	R 0.5	Z 0.025		(RAMP ON - RAISE Z 1/4 PITCH	)
G2		I 0	J 1.0601	Z 0.1			(FULL CIRCLE - RAISE Z 1 PITCH	)
G3		X -0.5	Y -1.5601	R 0.5	Z 0.025		(RAMP OFF - RAISE Z 1/4 PITCH	)
G1 G40		X 0	Y -2.0601				(RETURN TO HOME PLATE	)
M99							(RETURN TO PROGRAM	)
%								

APPLIED MATHEMATICS

     Advanced CNC Mill Programming and Applied Mathematics Level 2

Chapter 5
Geometry

Objectives

1. The student will demonstrate knowledge of the importance of geometry as it applies to CNC programming.

2. The student will demonstrate knowledge of the basic geometry.

3. The student will demonstrate knowledge of the basic geometry terminology.

4. The student will demonstrate knowledge of the how to use geometry to set up the problem.

    Advanced CNC Mill Programming and Applied Mathematics Level 2

Why Right Angle Trig & Geometry?

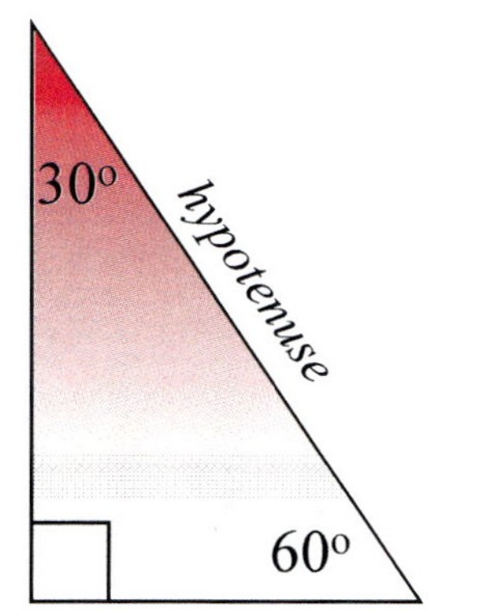

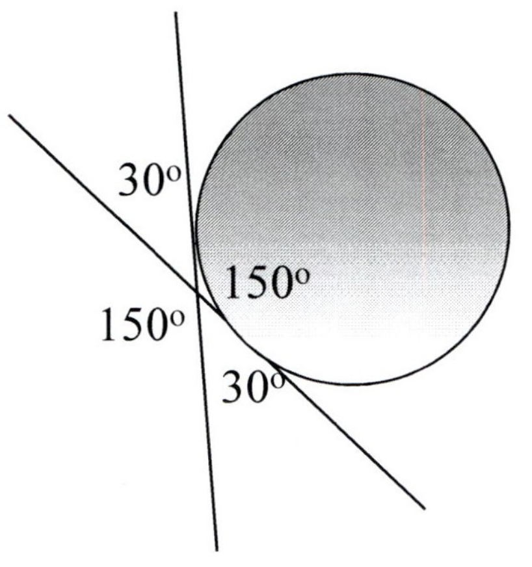

- A fundamental understanding of geometry and right angle trigonometry is essential to manually program CNC machines because print dimensions must be converted to a language CNC machines can understand—Cartesian XYZ coordinates

- The principles demonstrated here can be applied in many areas of manufacturing, such as Milling, Turning, RAM and Wire EDM, Punch, Laser, Waterjet, Grinding, and Inspection.

- Right angle Trig is used commonly for the following reasons:
  1. **Part Geometry**: To calculate toolpath end points when linear interpolating angled walls.
  2. **Circular Interpolation**: To calculate toolpath end and center points of arcs when circular interpolating.
  3. **Tool Tangents**: To offset the part shape by the tool radius when cutter diameter is not used.

    Advanced CNC Mill Programming and Applied Mathematics Level 2

1. PART GEOMETRY: Calculating angled walls.

International standards for dimensioning dictate that prints be drawn with the fewest dimensions possible, while completely defining the part. In the early days *over-dimensioning* was forbidden to avoid conflicts. Today *over-dimensioning* is prohibited due to the constraints of *sketcher*-based solid modeling. CNC programmers apply right angle trig to calculate the legs of triangles to convert missing dimensions to Cartesian coordinates for CNC programming.

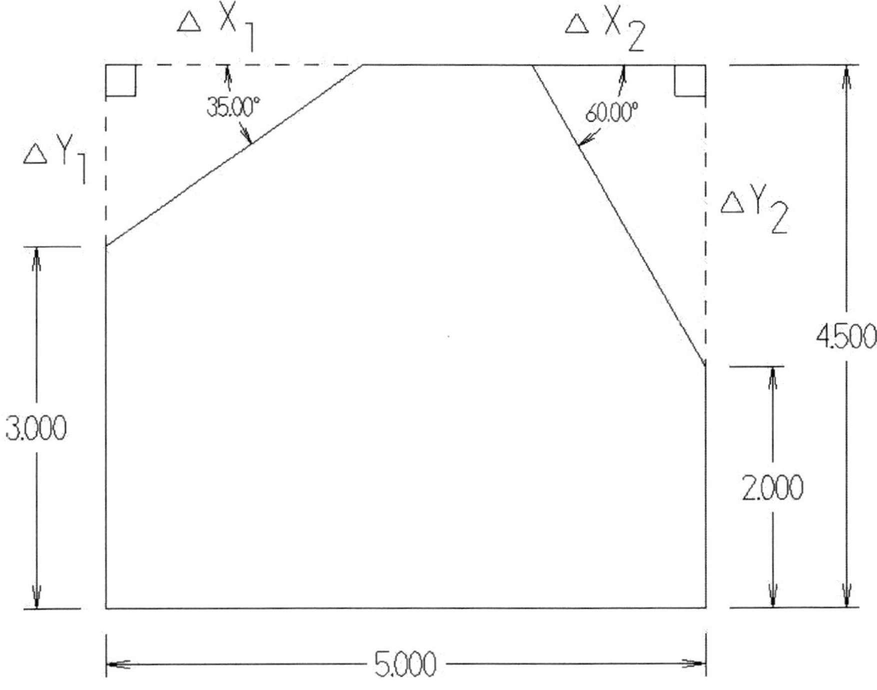

    Advanced CNC Mill Programming and Applied Mathematics Level 2

2. CIRCULAR INTERPOLATION: Calculating end and center points.

Trig is applied to calculate the end and center points of arcs and circles when circular interpolating. XY coordinates define the end point of an arc or circle, while IJ coordinates define the center point.

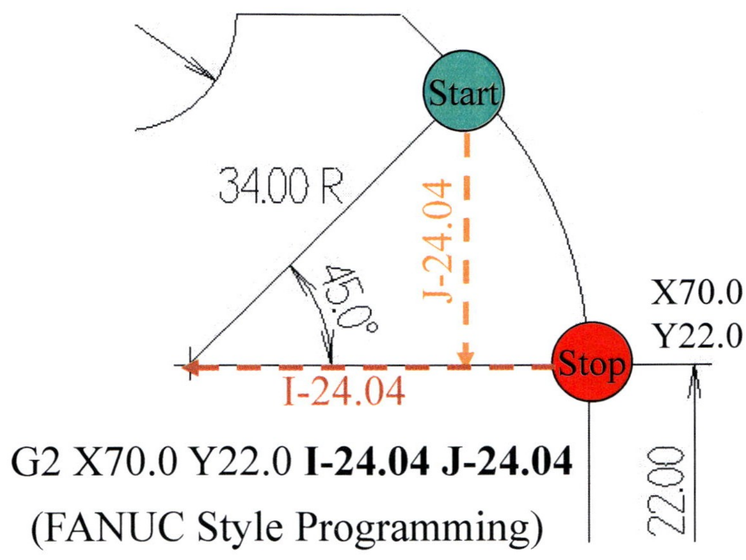

NOTE: There are two main styles of Circular Interpolation with respect to IJ coordinates:
1. FANUC: IJ coordinates are measured incrementally from the start point to the center.
2. Non-FANUC: IJ coordinates define the absolute position of the arc center.

    Advanced CNC Mill Programming and Applied Mathematics Level 2

3. TOOL TANGENTS: Offsetting geometry when CC is not used.

Right Angle Trig and Geometry are applied to offset your part shape when producing toolpaths without Cutter Compensation (CC) (mill) and Tool Nose Radius (TNR) (lathe). Toolpaths are offset by an amount equal to the tool radius, and the start/stop points are adjusted for their tangent points. Although Cutter Compensation performs these calculations automatically, not all machines use CC.

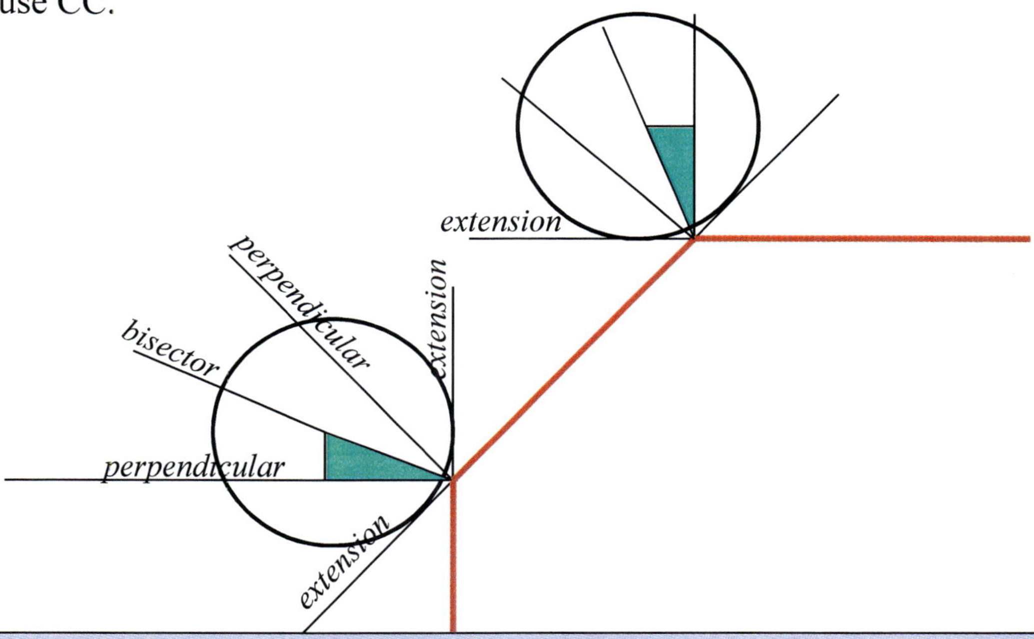

NOTE: Because of the extra work involved, and increased chance of error, it is strongly encouraged to use Cutter Compensation.

Geometry Terminology

Before properly demonstrating right angle trig, a fundamental understanding of *geometry* is necessary to set up your right triangles.

This section demonstrates ***applied geometry***. A comprehensive understanding of geometry is better served in a more thorough Geometry class.

Let's start with some simple definitions…

     Advanced CNC Mill Programming and Applied Mathematics Level 2

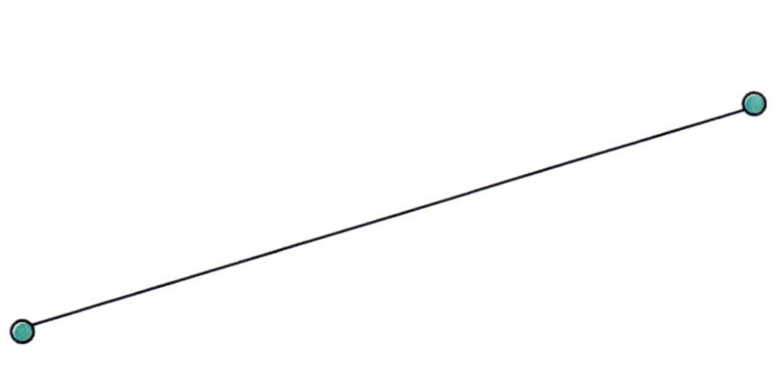

A straight **line** is the shortest distance between two points.

Two intersecting lines separated by 90 degrees are said to be at **right angles** from one another.

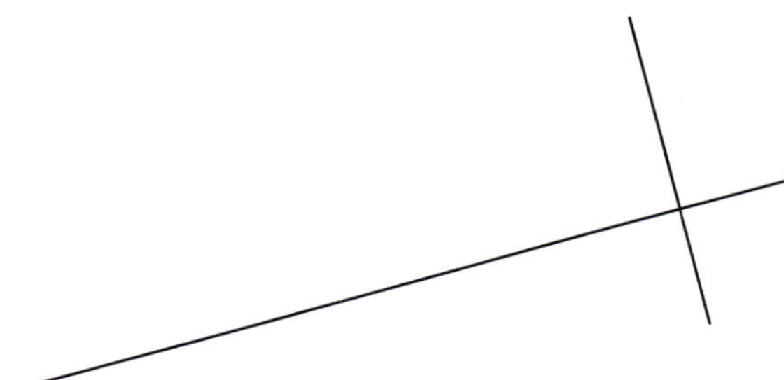

A **perpendicular** line is one that extends 90 degrees from a given line in either direction. "*Normal*" is synonymous with perpendicular. Both terms are commonly used in CAD.

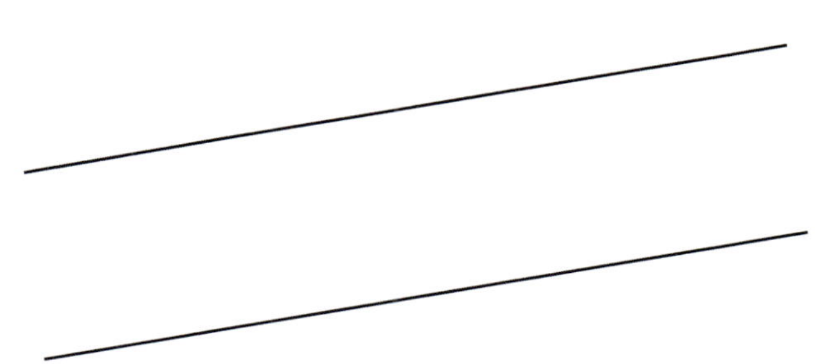

Two lines are **parallel** if when extended infinitely they never intersect with each other. "*Offset*" is synonymous with parallel.

   Advanced CNC Mill Programming and Applied Mathematics Level 2

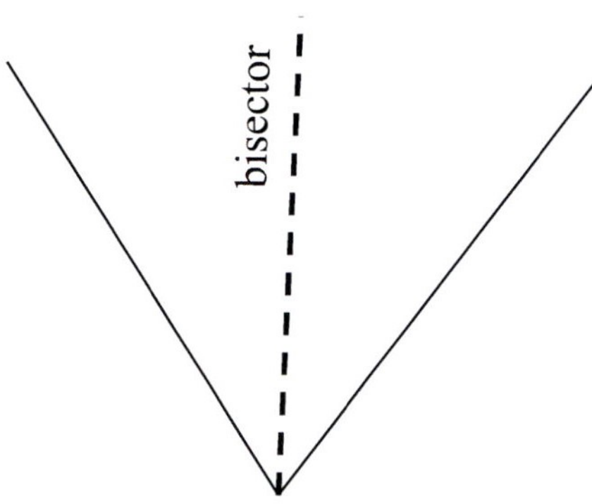

A **bisector** line is a position halfway between two lines.

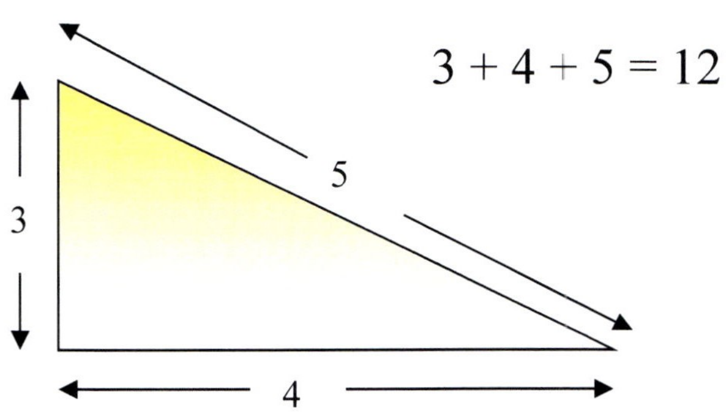

The **perimeter** of any shape is the sum of the lengths of the sides.

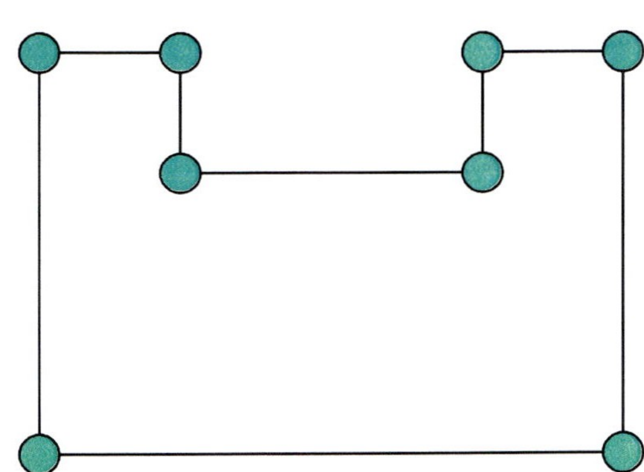

A **vertex** is a corner point of a geometric entity.

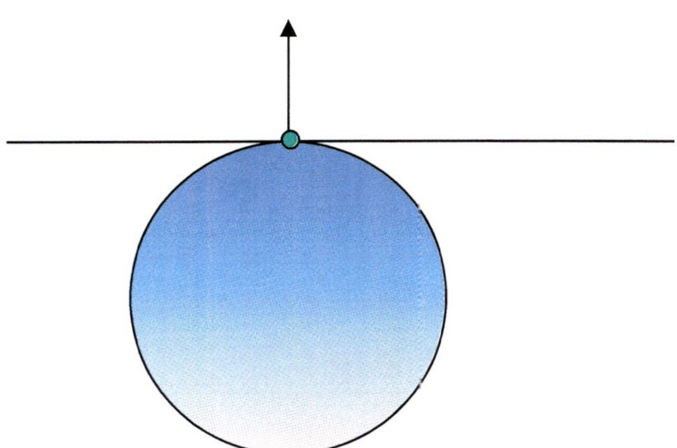

Tangent is a condition where two geometric entities touch and their *normals* match.

     Advanced CNC Mill Programming and Applied Mathematics Level 2

Diameter = Radius x 2

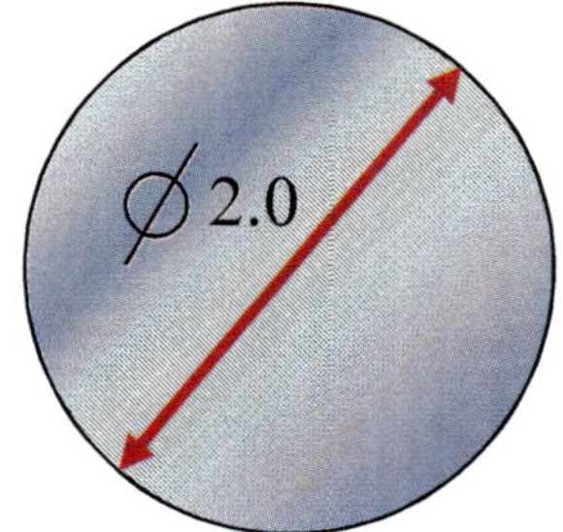

The **diameter** of a circle is a straight line and is measured from one side of the circle to the other, passing through the circle center.

Circumference = π x Diameter

Area of a circle = π r²

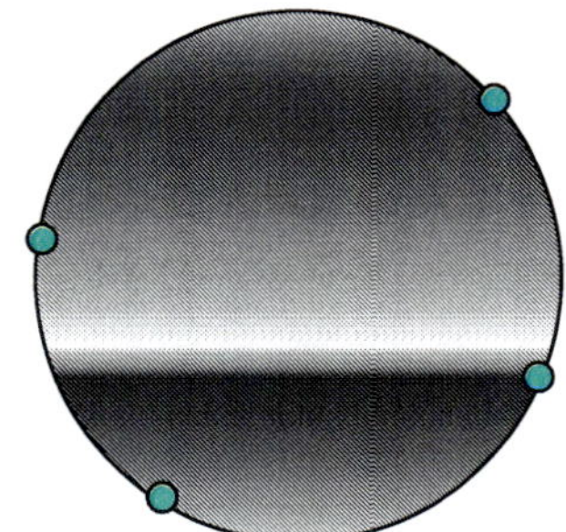

The **circumference** of a circle is any point on the outer edge of the circle.

Note: π = *Pi*

Radius = Diameter / 2

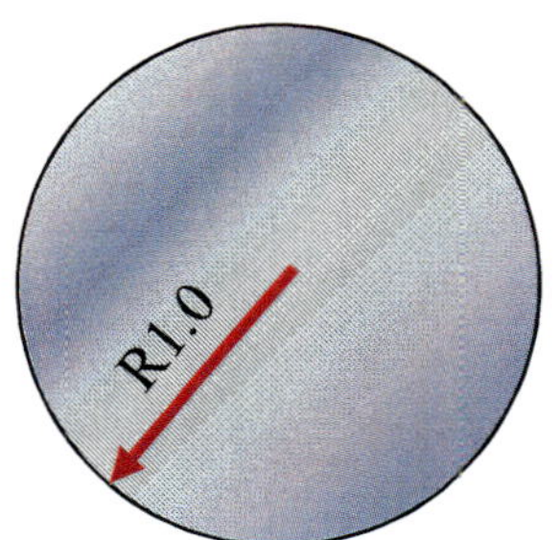

The **radius** of the circle is equal to the ½ the diameter. It's measured from the center of the arc or circle to any point on the circumference.

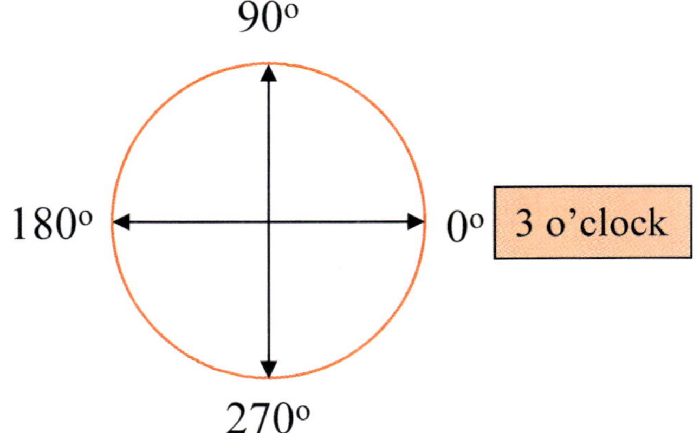

Angles are measured counterclockwise from the 3 o'clock.

     Advanced CNC Mill Programming and Applied Mathematics Level 2

53

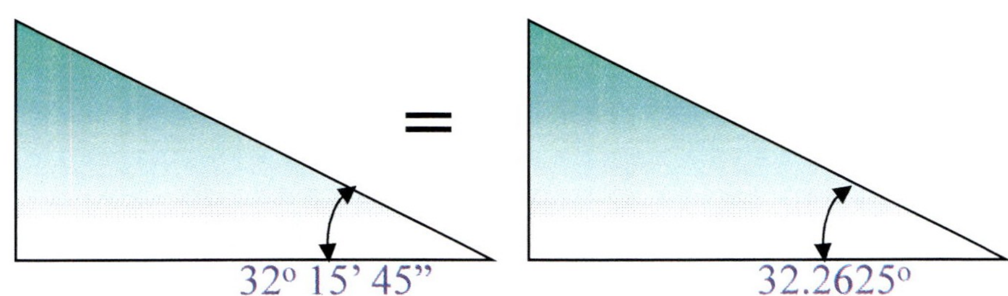

Angles are measured *radially* between two lines.

Angles are measured in degrees-minutes-seconds, or decimal degrees.

The symbols are placed in the right corner.
For example, 32° 15' 45".

-A **degree** is 1/360th of a circle.

-A **minute** is 1/60th of a degree.

-A **second** is 1/60th of a minute.

Degrees-Minutes-Seconds must be converted to decimal degrees when using trigonometry.

① **Convert minutes to degrees:**

Degrees = Minutes / 60
0.25 deg = 15 / 60

② **Convert seconds to degrees:**

Degrees = Seconds / 60 / 60
Degrees = Seconds / 3600
0.0125 = 45 / 3600

③ **Add:** 32.0000
+ 0.2500
0.0125
______
32.2625 degrees

   Advanced CNC Mill Programming and Applied Mathematics Level 2

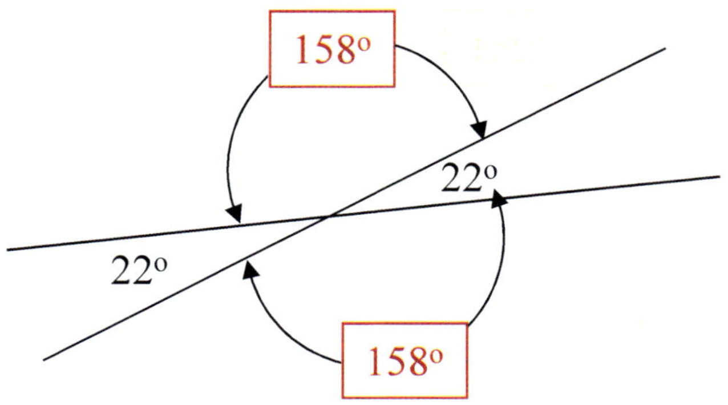

Where two lines intersect, the opposite angles are equal.

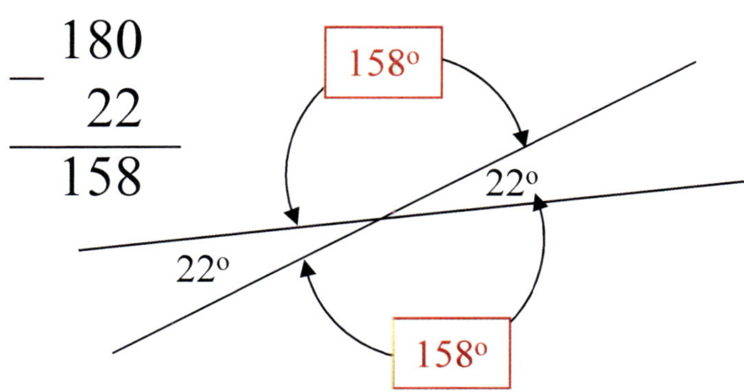

$$180 - 22 = 158$$

The **compliment** is found by subtracting the known angle from 180º.

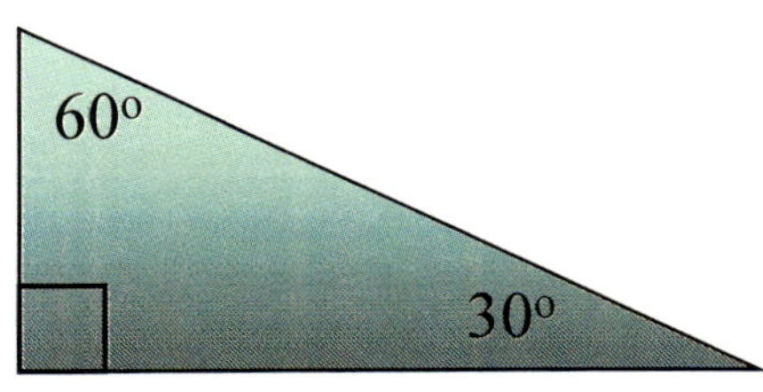

The sum of the angles of any triangle equals 180º. Therefore, the sum of the angles opposite the right angle equals 90º.

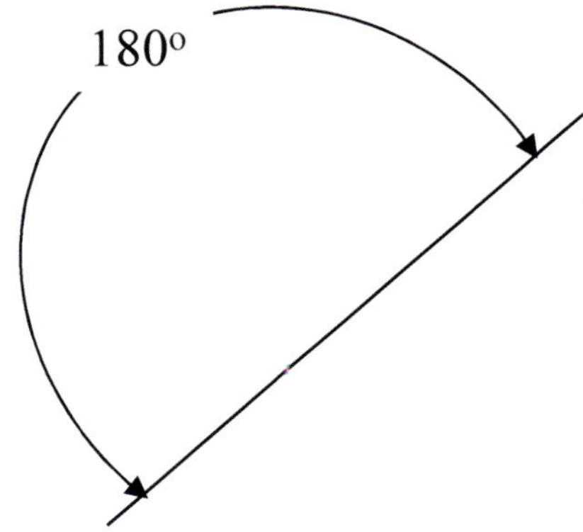

Line angles measure 180 degrees.

     Advanced CNC Mill Programming and Applied Mathematics Level 2

Chapter 6
Right Angle Trigonometry

     Advanced CNC Mill Programming and Applied Mathematics Level 2

Objectives

1. The student will demonstrate knowledge of the importance of right angle trigonometry as it applies to CNC programming.

2. The student will demonstrate knowledge basic right angle trigonometry.

3. The student will demonstrate knowledge basic trigonometry terminology.

4. The student will demonstrate how to use right angle trigonometry to set up the problem.

Trigonometry

With a fundamental understanding of geometry, now we can better demonstrate *trigonometry*. Although many principles of trig apply, right angle trig is the most common form used when programming CNC machines because everything must relate to the horizontal and vertical axes, more commonly referred to as X and Y-axis.

This section demonstrates ***applied right angle trig***. A comprehensive understanding of trigonometry is better served in a more thorough Trigonometry class.

    Advanced CNC Mill Programming and Applied Mathematics Level 2

Trigonometry Definitions

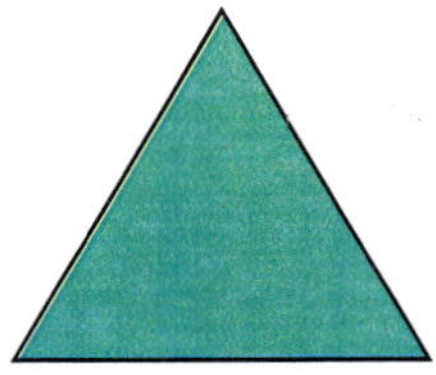

A **triangle** is made up of three lines.

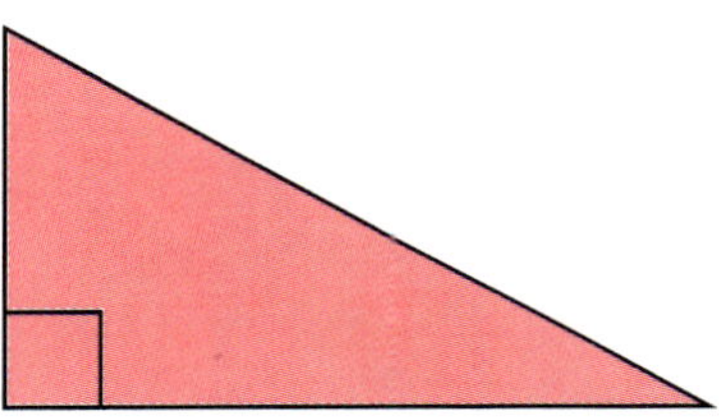

A **right triangle** is a triangle where two of the three legs form a 90° angle.

The **Side Opposite** is defined as the leg opposite the *known* angle.

The **Side Adjacent** is defined as the leg *adjacent*—or next to—the *known* angle.

The **Hypotenuse** is the side opposite the right angle. It's the longest side.

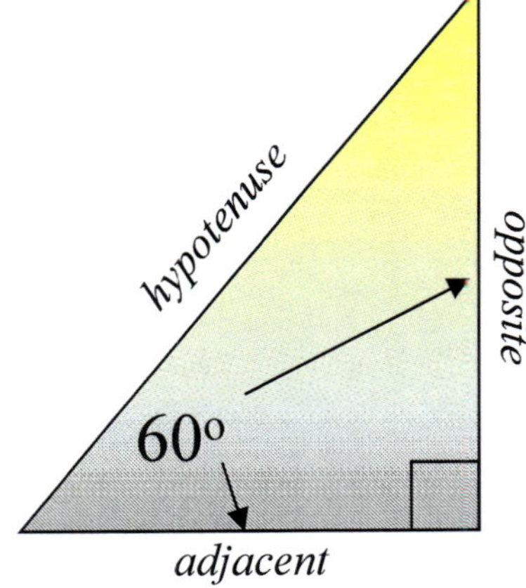

Pythagorean Theorem

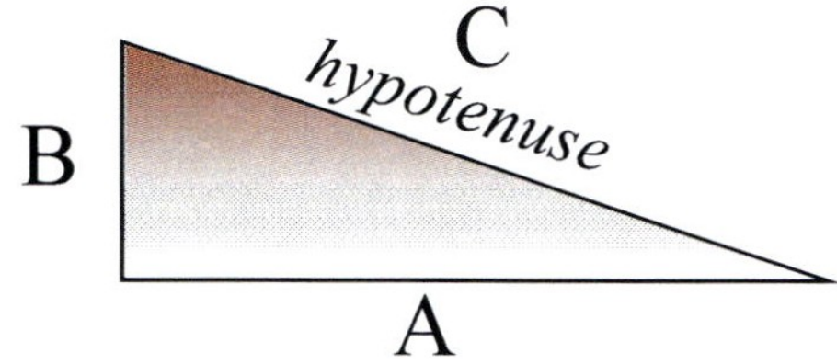

Given the following **Pythagorean Theorem**, the square of the hypotenuse is equal to the sum of the squares of the other two sides:

$$C^2 = A^2 + B^2 \qquad or... \qquad C = \sqrt{A^2 + B^2}$$

Using *Algebra* you can **transpose** to solve for any side:

$$A^2 = C^2 - B^2 \qquad or... \qquad A = \sqrt{C^2 - B^2}$$

$$B^2 = C^2 - A^2 \qquad or... \qquad B = \sqrt{C^2 - A^2}$$

Pythagorean Theorem

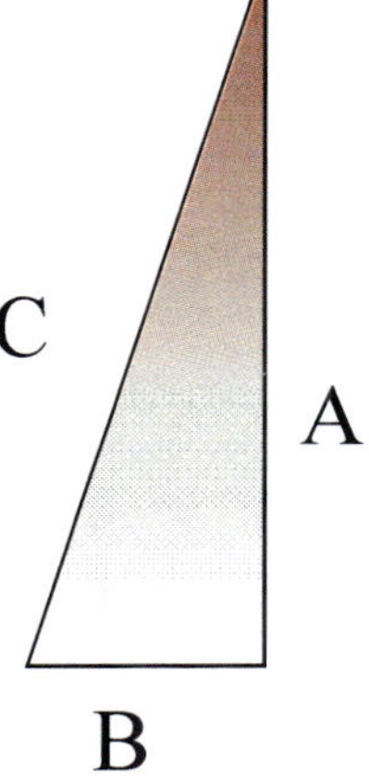

$$C = \sqrt{A^2 + B^2}$$
$$C = \sqrt{4^2 + 3^2}$$
$$C = \sqrt{16 + 9}$$
$$C = \sqrt{25}$$
$$C = 5$$

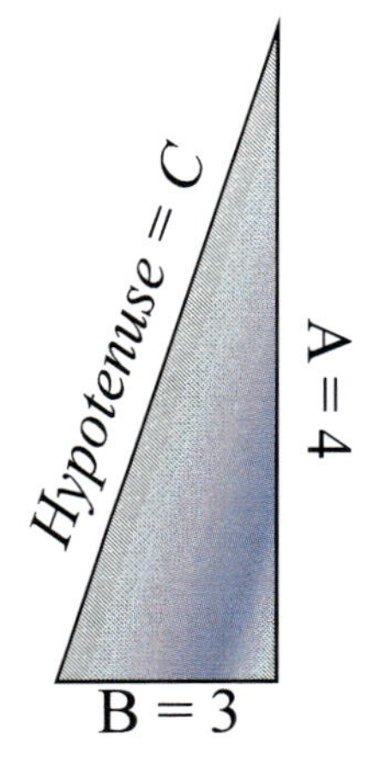

$$A = \sqrt{C^2 - B^2}$$
$$A = \sqrt{5^2 - 3^2}$$
$$A = \sqrt{25 - 9}$$
$$A = \sqrt{16}$$
$$A = 4$$

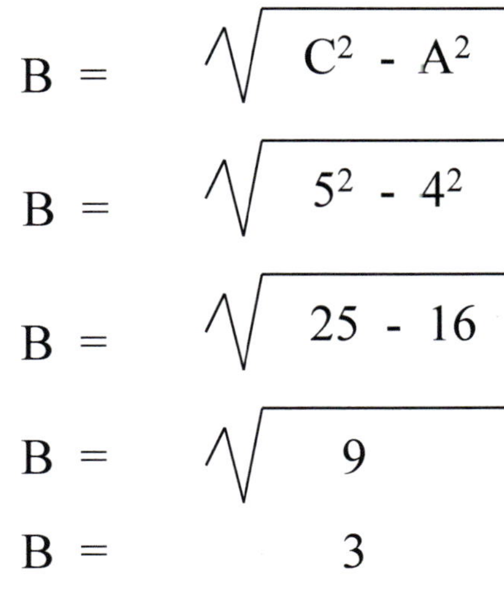

$$B = \sqrt{C^2 - A^2}$$
$$B = \sqrt{5^2 - 4^2}$$
$$B = \sqrt{25 - 16}$$
$$B = \sqrt{9}$$
$$B = 3$$

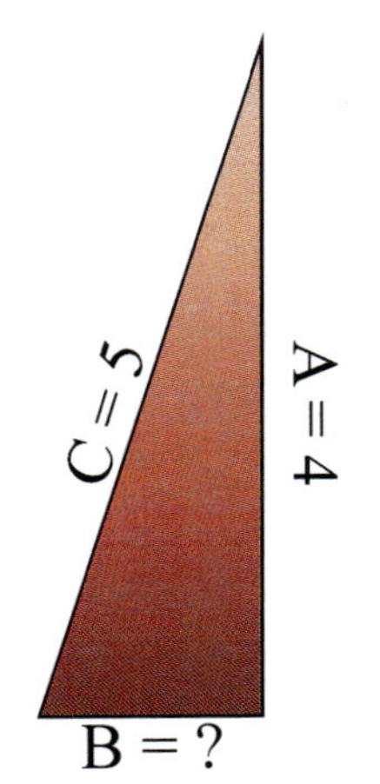

     Advanced CNC Mill Programming and Applied Mathematics Level 2

Trigonometric Functions
For Finding Approximate Side Lengths

$$\sin \theta = \frac{opposite}{hypotenuse}$$

$$\cos \theta = \frac{adjacent}{hypotenuse}$$

$$\tan \theta = \frac{opposite}{adjacent} = \frac{rise}{run}$$

$$\sin \theta = \frac{opposite}{hypotenuse}$$

$$\sin(60) = \frac{opposite}{1.375}$$

$$0.866 = \frac{opposite}{1.375}$$

$opposite = 0.866 \times 1.375$

$opposite = 1.19078$

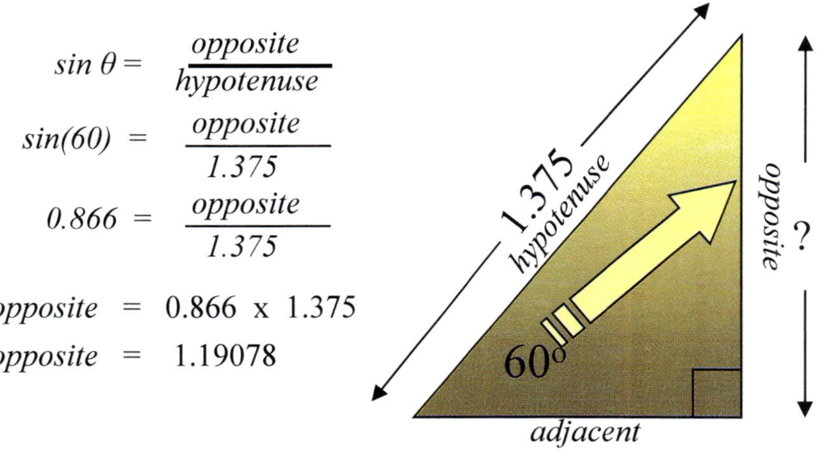

$$\cos \theta = \frac{adjacent}{hypotenuse}$$

$$\cos(60) = \frac{adjacent}{1.375}$$

$$0.500 = \frac{adjacent}{1.375}$$

$adjacent = 0.500 \times 1.375$

$adjacent = 0.6875$

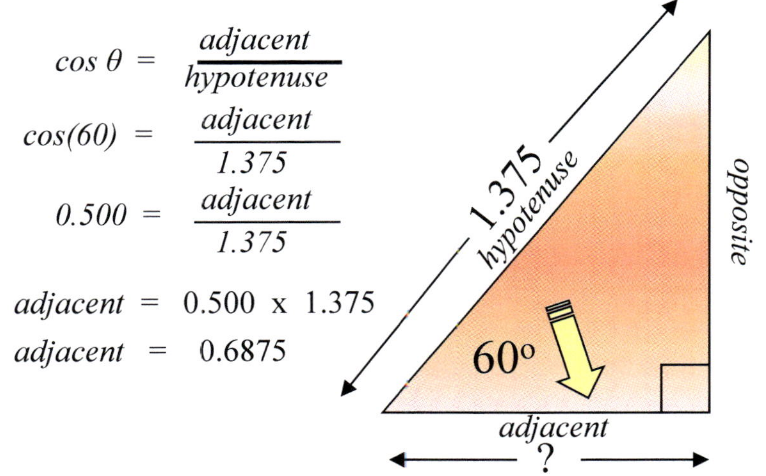

Finding **Opposite** when *adjacent* & *angle* known:

$$\tan \theta = \frac{opposite}{adjacent}$$

$$\tan(60) = \frac{opposite}{.6875}$$

$$1.732 = \frac{opposite}{.6875}$$

$opposite = 1.732 \times .6875$

$opposite = 1.19078$

Finding **adjacent** when *opposite* & *angle* known:

$$\tan \theta = \frac{opposite}{adjacent}$$

$$\tan(60) = \frac{1.19078}{adjacent}$$

$$1.732 = \frac{1.19078}{adjacent}$$

$$adjacent = \frac{1.19078}{1.732}$$

$adjacent = 0.6875$

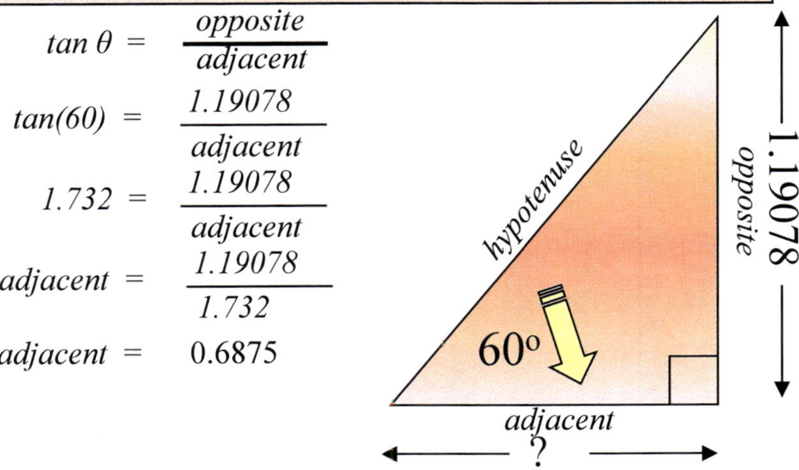

    Advanced CNC Mill Programming and Applied Mathematics Level 2

Trigonometric Functions
For Finding Approximate Angles (θ)

$$sin^{-1} = \frac{opposite}{hypotenuse}$$

$$cos^{-1} = \frac{adjacent}{hypotenuse}$$

$$tan^{-1} = \frac{opposite}{adjacent} = \frac{rise}{run}$$

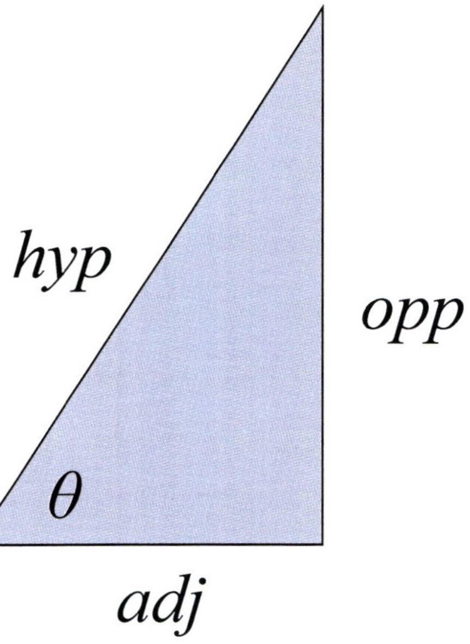

$$\theta = \sin^{-1} \frac{opposite}{hypotenuse}$$

$$\theta = \sin^{-1} \frac{1.5}{1.65}$$

$$\theta = \quad 65.38 \text{ degrees}$$

$$\theta = \cos^{-1} \frac{adjacent}{hypotenuse}$$

$$\theta = \cos^{-1} \frac{0.6875}{1.65}$$

$$\theta = \quad 65.38 \text{ degrees}$$

$$\theta = \tan^{-1} \frac{opposite}{adjacent}$$

$$\theta = \tan^{-1} \frac{1.500}{0.6875}$$

$$\theta = \quad 65.38 \text{ degrees}$$

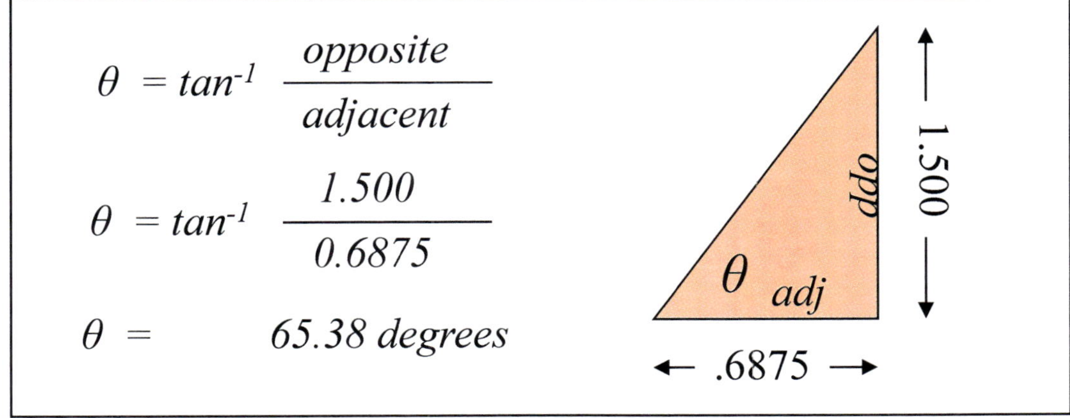

   Advanced CNC Mill Programming and Applied Mathematics Level 2

Converting Polar Coordinates to Rectangular Coordinates

$$X = \text{r} \; x \; \cos(\theta)$$
$$= 3.0 \; x \; 0.5$$
$$= 1.5$$

$$Y = \text{r} \; x \; \sin(\theta)$$
$$= 3.0 \; x \; 0.866$$
$$= 2.5981$$

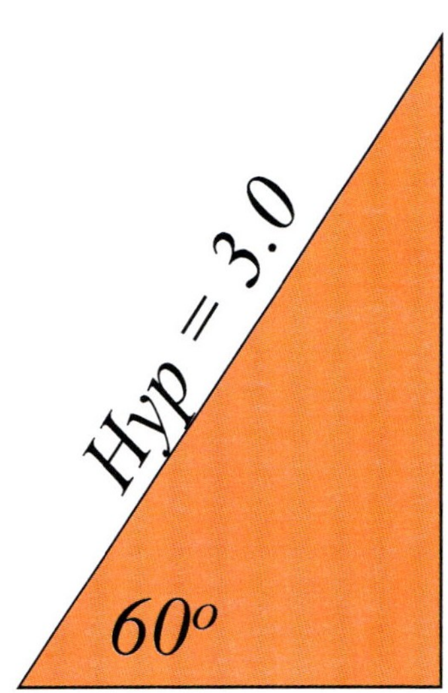

CNC Program 1
Solving Right Triangles For Linear Interpolation

The part below is two dimensional in nature and requires simple right angle trig calculations to calculate the missing XY coordinate values.
This part will be held on the table by four bolts as shown. It will lie on a sacrificial riser so the cutter can climb mill the perimeter while machining 0.010" below the bottom of the part—into the riser.   Z0 will be located at the top of the part.

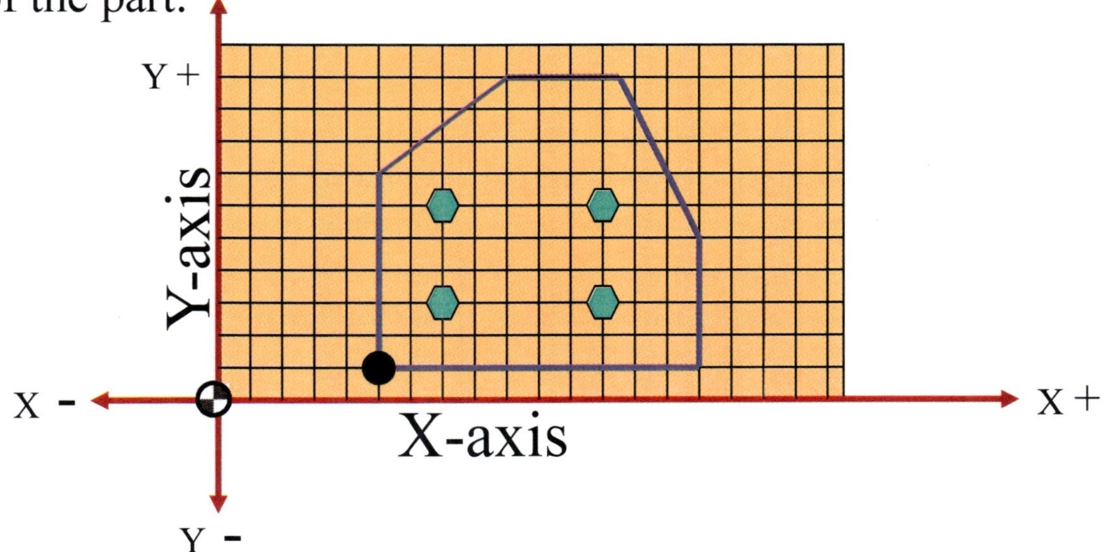

CNC Program 1

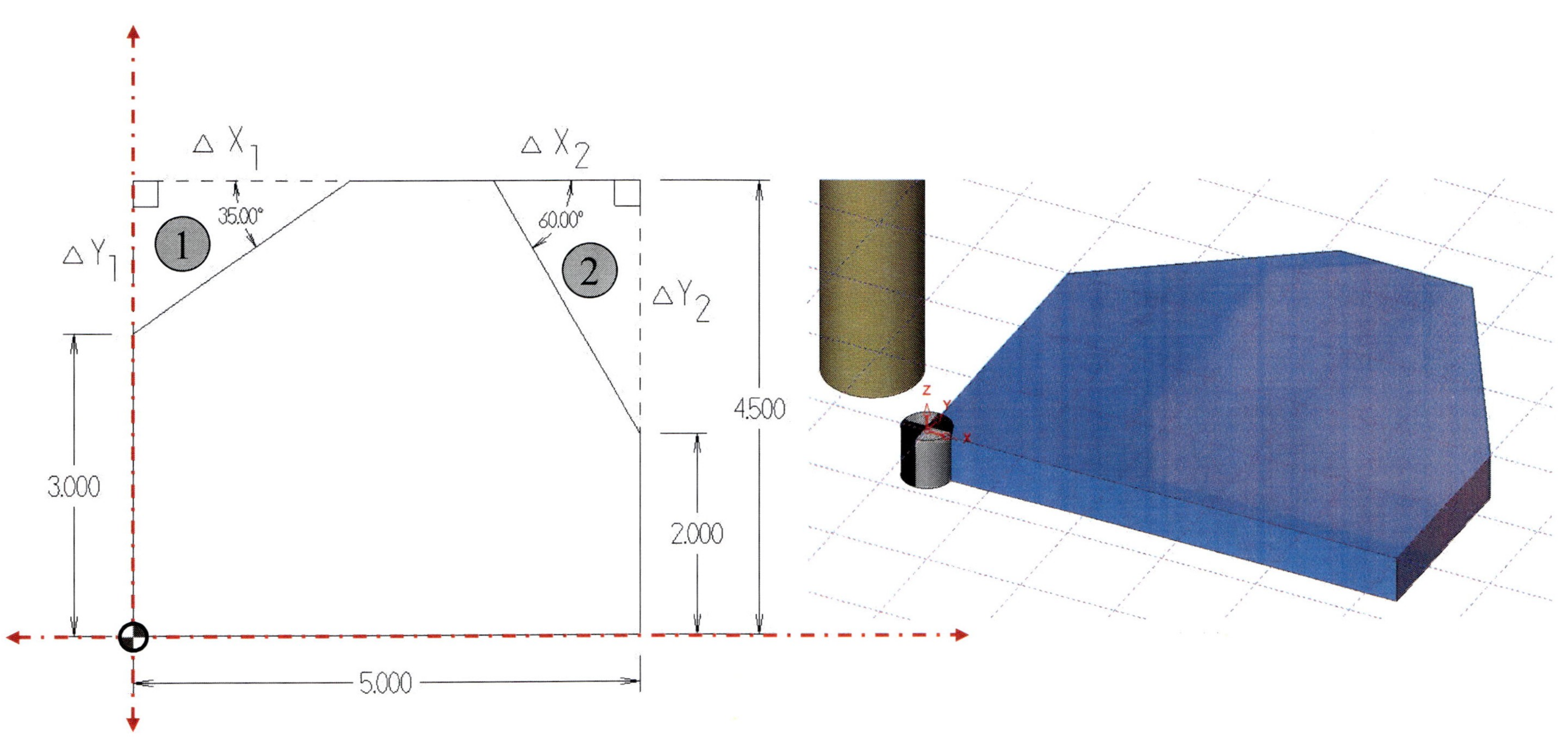

     Advanced CNC Mill Programming and Applied Mathematics Level 2

Solving For the Side Lengths:

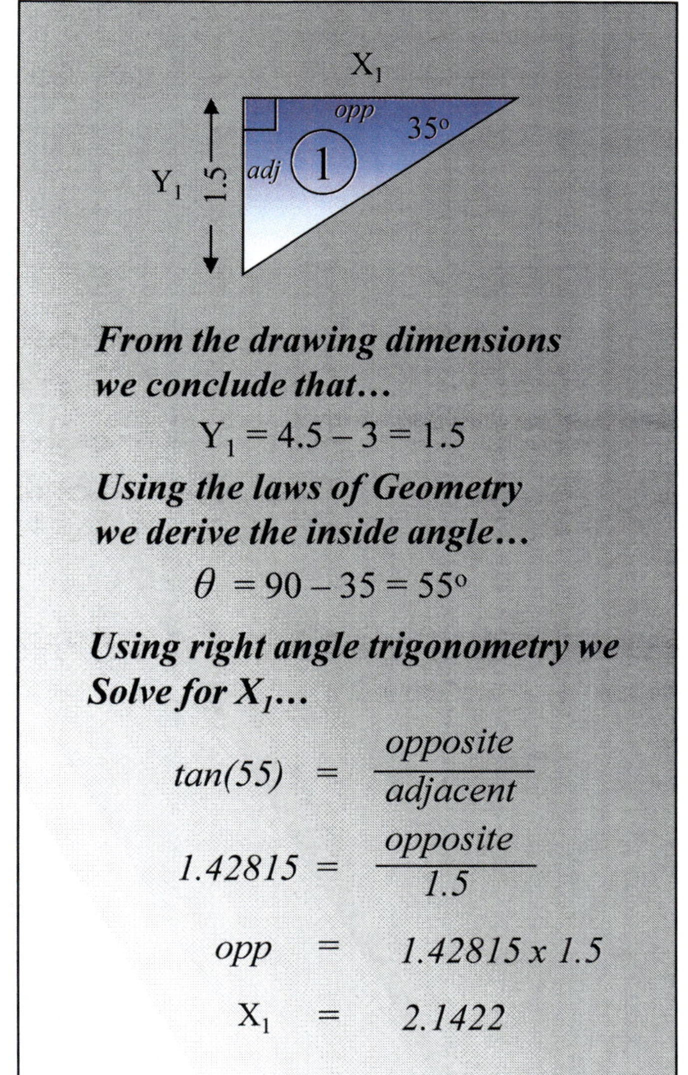

From the drawing dimensions we conclude that…

$$Y_1 = 4.5 - 3 = 1.5$$

Using the laws of Geometry we derive the inside angle…

$$\theta = 90 - 35 = 55°$$

Using right angle trigonometry we Solve for X_1…

$$tan(55) = \frac{opposite}{adjacent}$$

$$1.42815 = \frac{opposite}{1.5}$$

$$opp = 1.42815 \times 1.5$$

$$X_1 = 2.1422$$

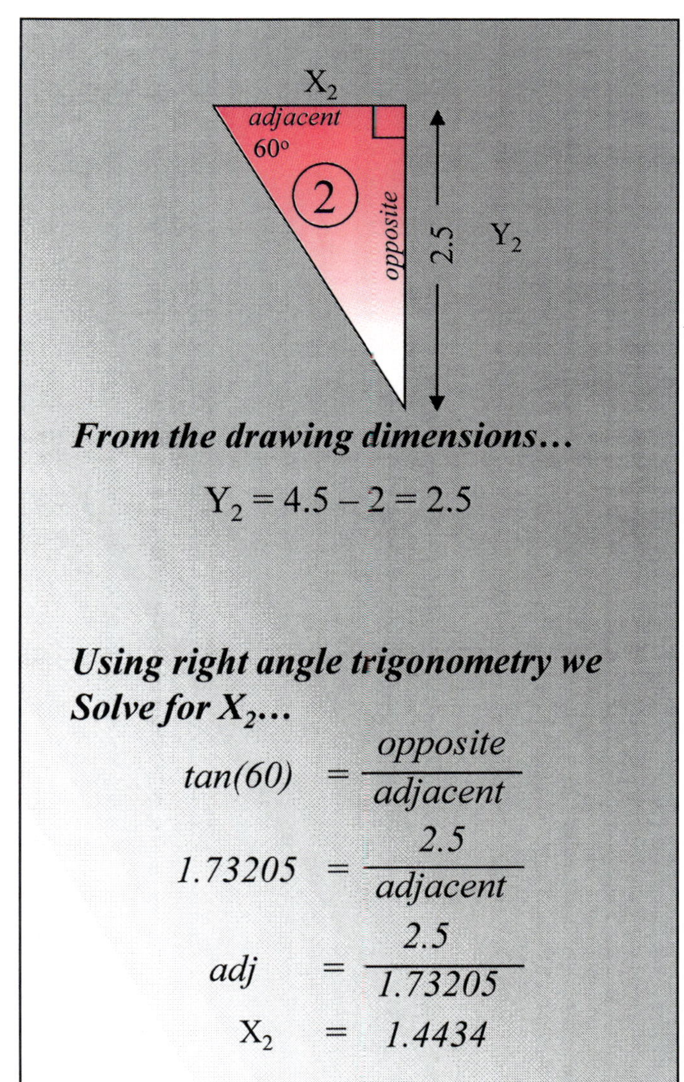

From the drawing dimensions…

$$Y_2 = 4.5 - 2 = 2.5$$

Using right angle trigonometry we Solve for X_2…

$$tan(60) = \frac{opposite}{adjacent}$$

$$1.73205 = \frac{2.5}{adjacent}$$

$$adj = \frac{2.5}{1.73205}$$

$$X_2 = 1.4434$$

Mapping Cartesian Coordinates

After solving for the missing values, write the XY coordinates at each corner.
These coordinates become the body of your CNC program.

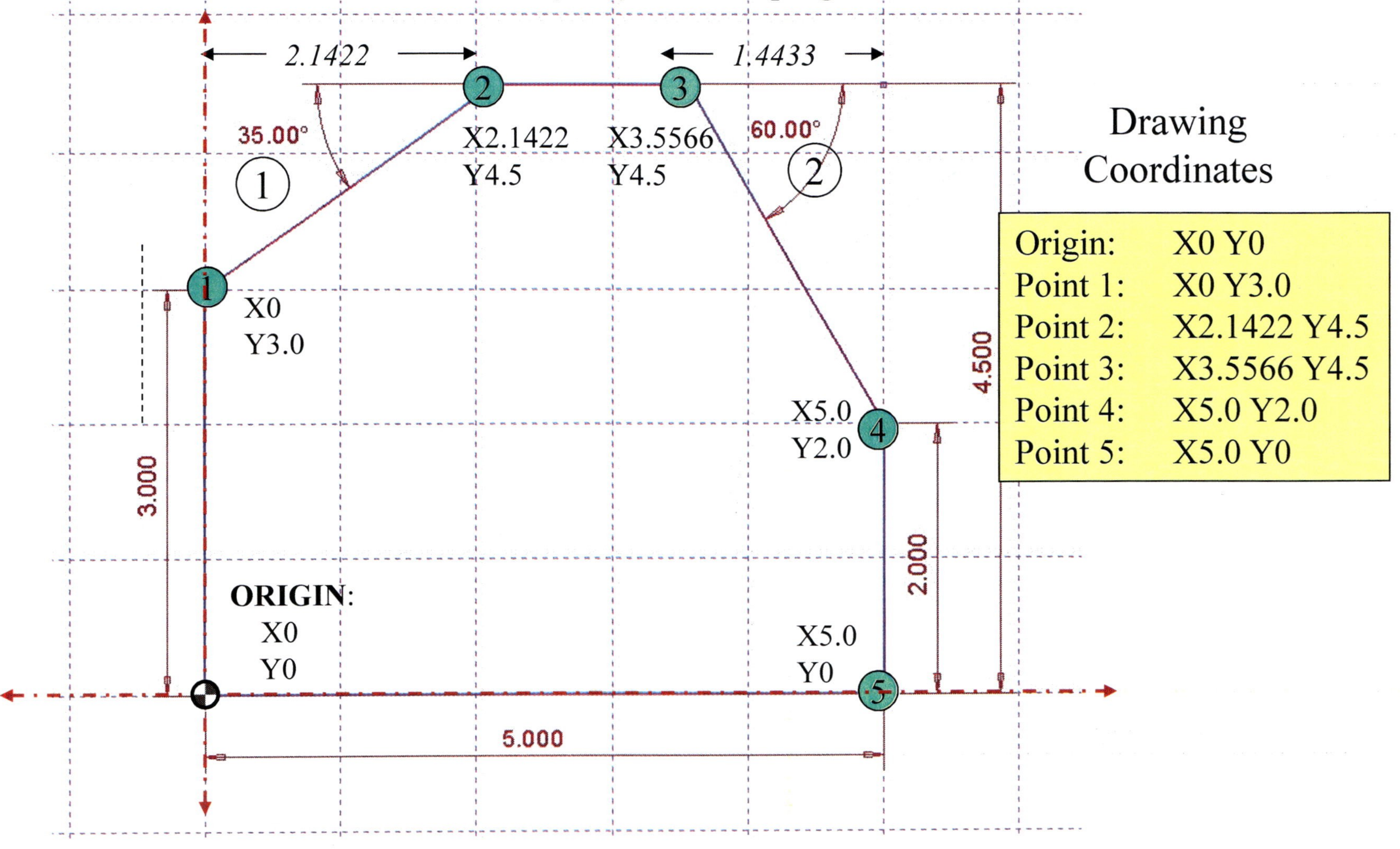

   Advanced CNC Mill Programming and Applied Mathematics Level 2

Drawing Coordinates Become the Body

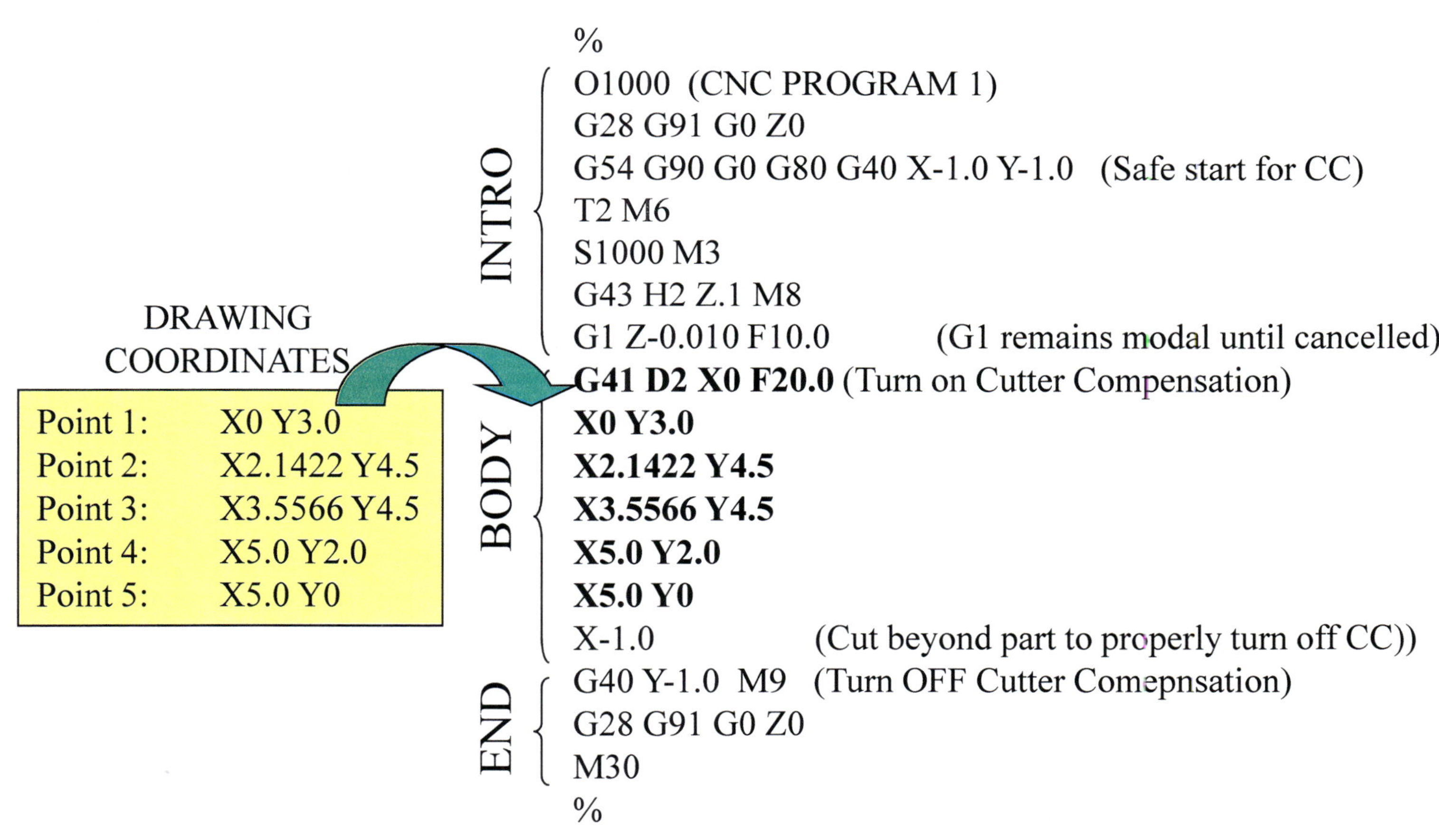

     Advanced CNC Mill Programming and Applied Mathematics Level 2

FANUC Style
Circular Interpolation Review

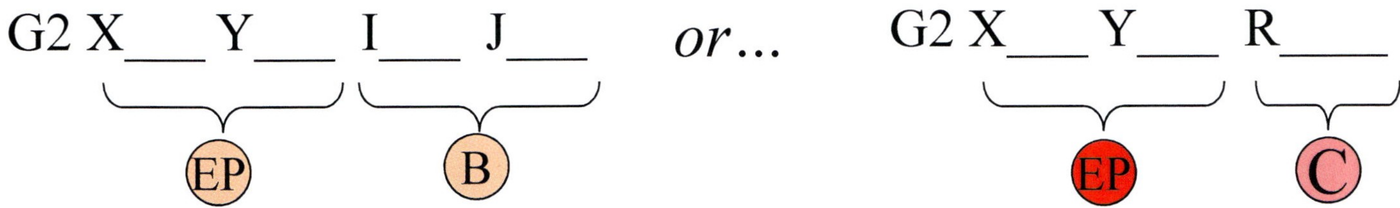

EP End Point (destination) of arc or circle.

B Incremental distances in X & Y respectively from Start Point (SP) to arc center.

C Radius – used instead of IJ. (Cannot be used on 360° circles.)
Minus sign (-) required on moves greater than 180°.

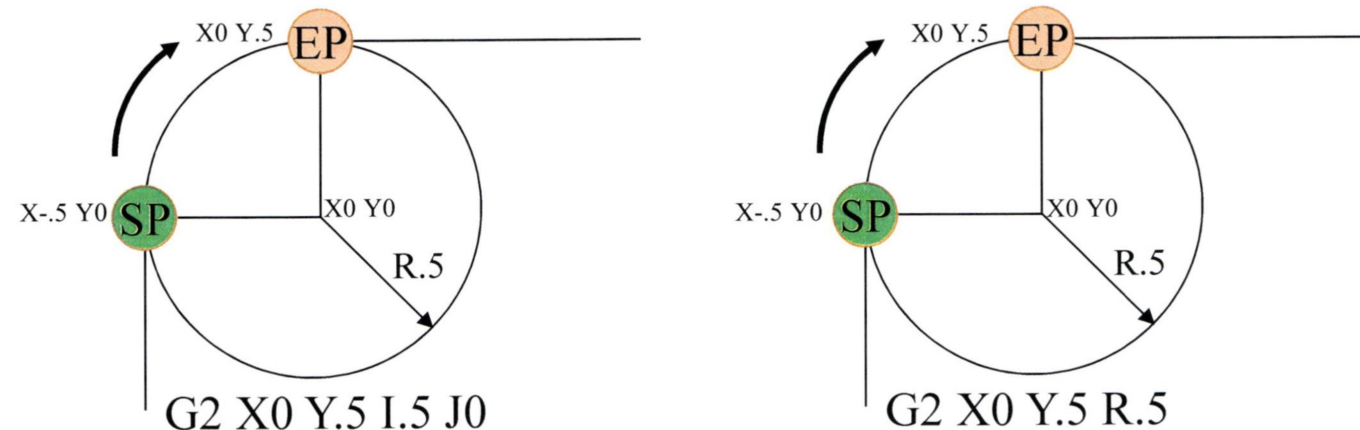

   Advanced CNC Mill Programming and Applied Mathematics Level 2

CNC Program 2
Solving Right Triangles For Circular Interpolation

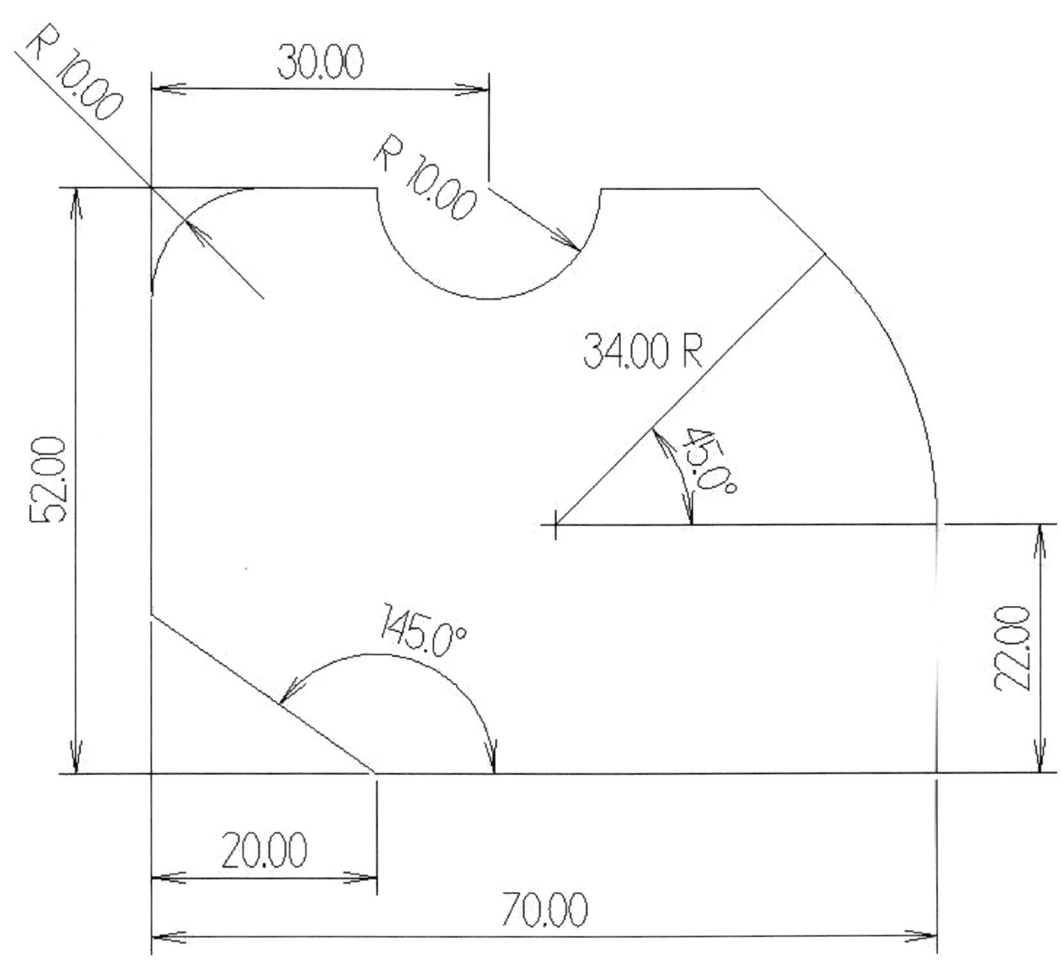

The markers below identify the start and stop points for each circular block.
Arcs 1 and 2 are easily solved because the arcs are tangent to horizontal and
vertical lines. Arc 3 requires right angle trig to solve. And the marked corner
point must be solved for using the principles of geometry and right angle trig.

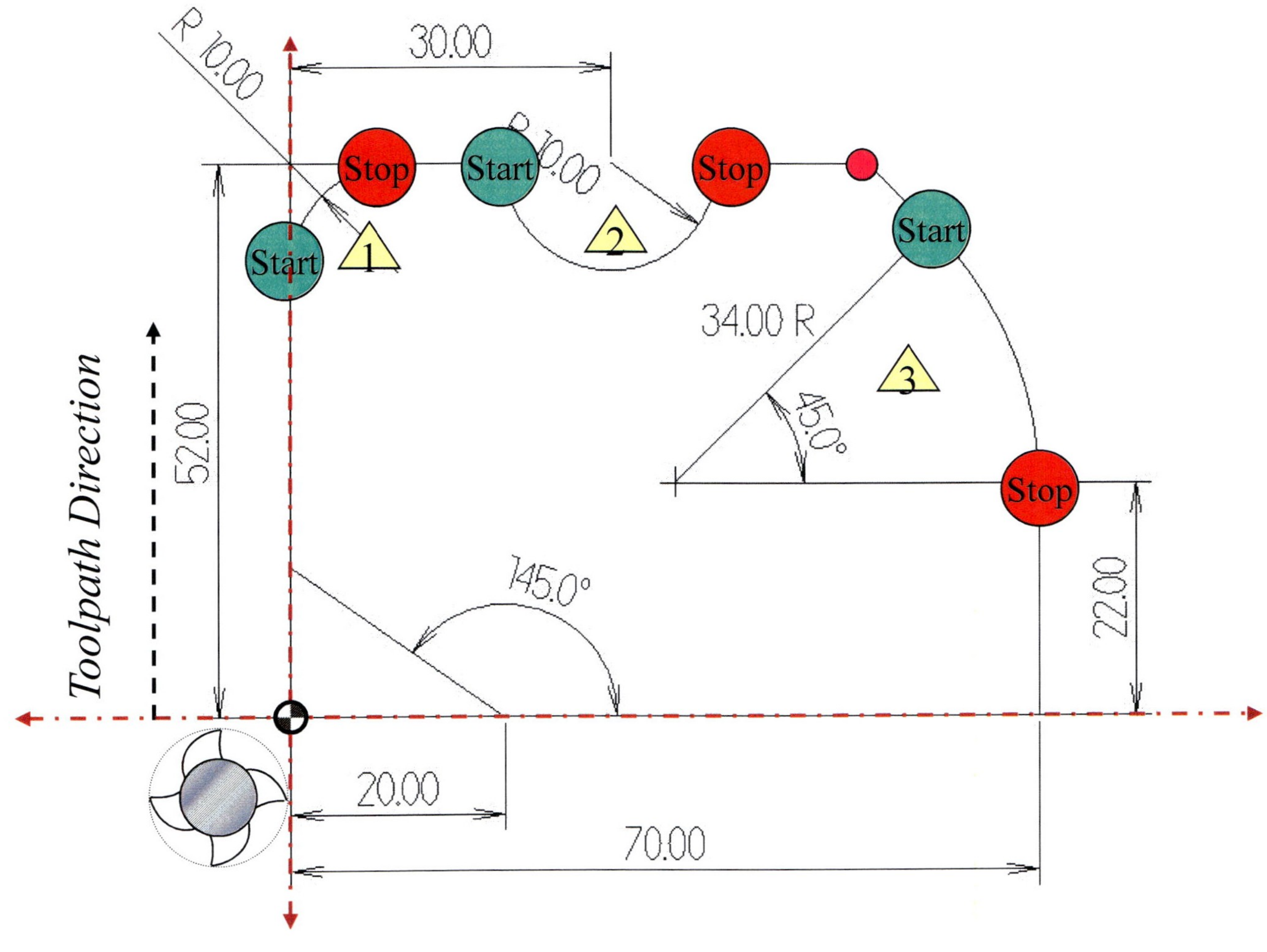

The radius of Arc 1 is 10 mm and the tangent lines are horizontal and vertical.
The distance from the start of the arc to the center—measured along the X-axis
is *positive* 10 mm, therefore I = 10.
The distance from the start of the arc to the center—measured along the Y-axis
is *zero*, therefore J = 0.

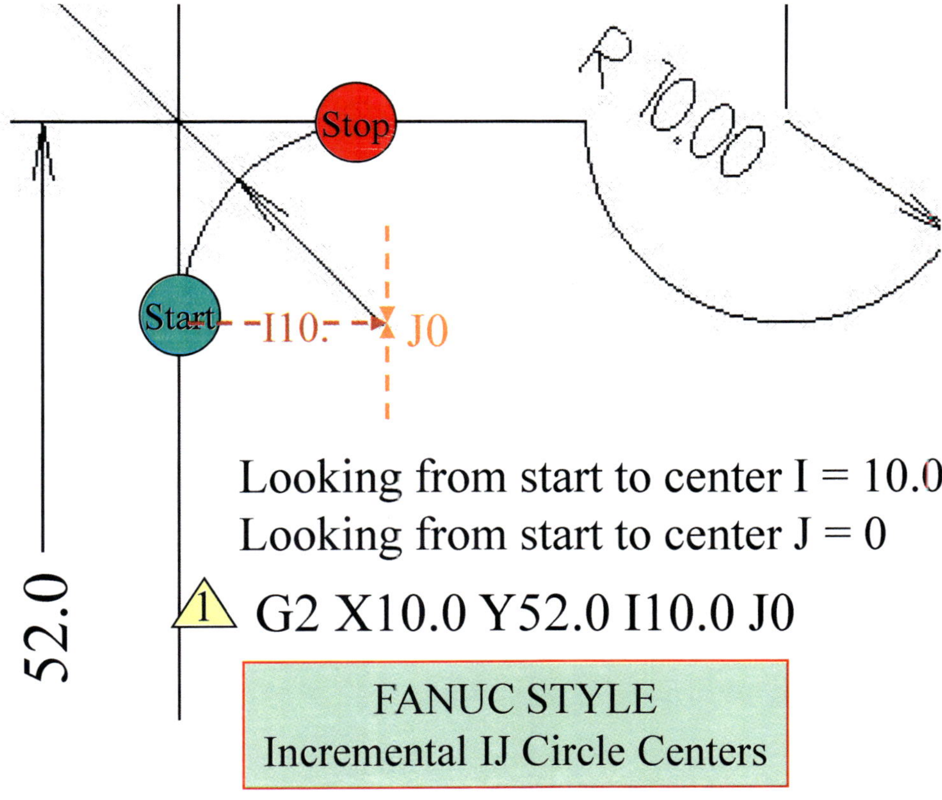

The radius of Arc 2 is 10 mm and both tangent lines are horizontal.
The distance from the start of the arc to the center—measured along the X-axis
is *positive* 10 mm, therefore I = 10.0
The distance from the start of the arc to the center—measured along the Y-axis
is *zero*, therefore J = 0.

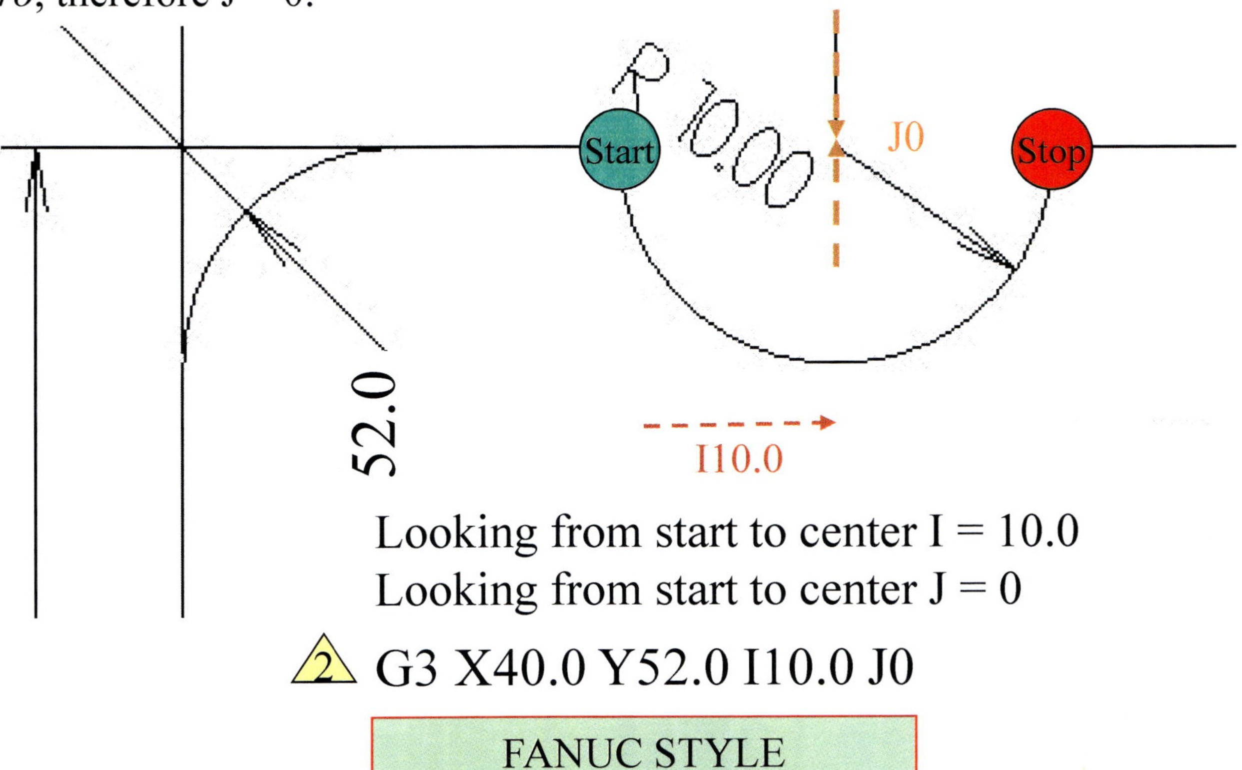

Looking from start to center I = 10.0
Looking from start to center J = 0

2. G3 X40.0 Y52.0 I10.0 J0

FANUC STYLE
Incremental IJ Circle Centers

     Advanced CNC Mill Programming and Applied Mathematics Level 2

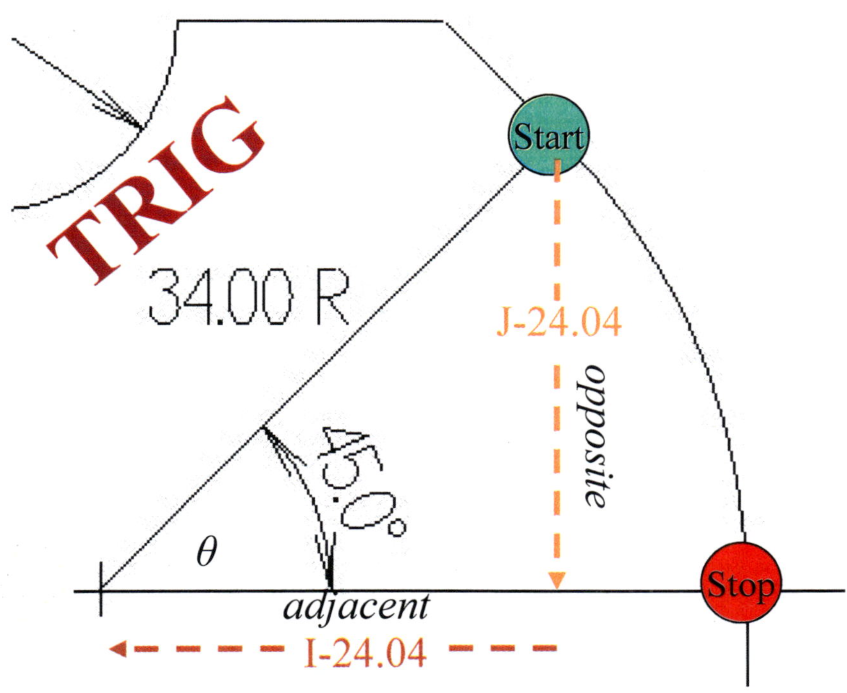

From start to center I = -24.04
From start to center J = -24.04

△3 G2 X70.0 Y22.0 I-24.04 J-24.04

Note: When $\theta = 45$, both legs are equal.

FANUC STYLE
Incremental IJ Circle Centers

The radius of Arc 3 is 34 mm and the tangent lines are vertical and angled.
Step 1: Construct a right triangle from the arc center to the tangent point at the top. Using the table below you can calculate the leg lengths. The legs are equal when the angle is 45°.

The distance from the start of the arc to the center—measured along the X-axis is *negative* 24.04 mm, therefore I = -24.04
The distance from the start of the arc to the center—measured along the Y-axis is *negative 24.04*, therefore J = -24.04

$$sin(45) = \frac{opposite}{hypotenuse}$$

$$0.7071 = \frac{opposite}{34}$$

$$opposite = 0.7071 \times 34$$

$$opposite = 24.04$$

Geometry & Trig For Linear Interpolation

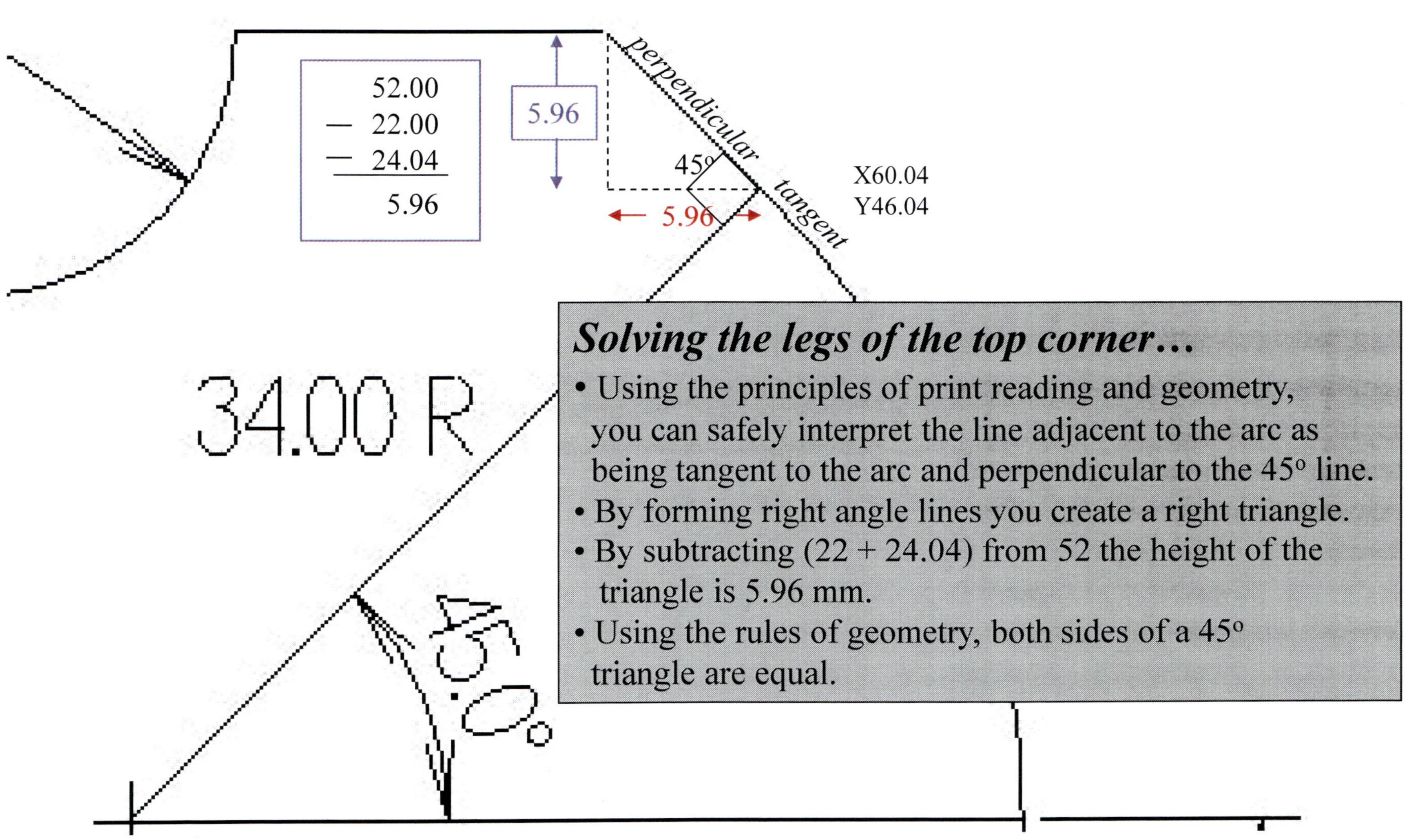

Solving the legs of the top corner…
- Using the principles of print reading and geometry, you can safely interpret the line adjacent to the arc as being tangent to the arc and perpendicular to the 45° line.
- By forming right angle lines you create a right triangle.
- By subtracting (22 + 24.04) from 52 the height of the triangle is 5.96 mm.
- Using the rules of geometry, both sides of a 45° triangle are equal.

    Advanced CNC Mill Programming and Applied Mathematics Level 2

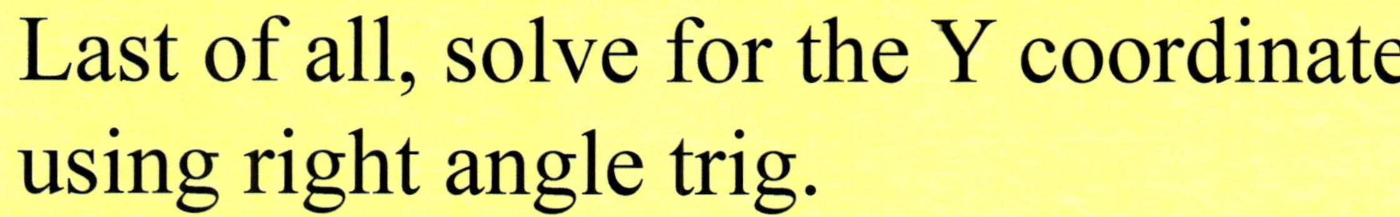

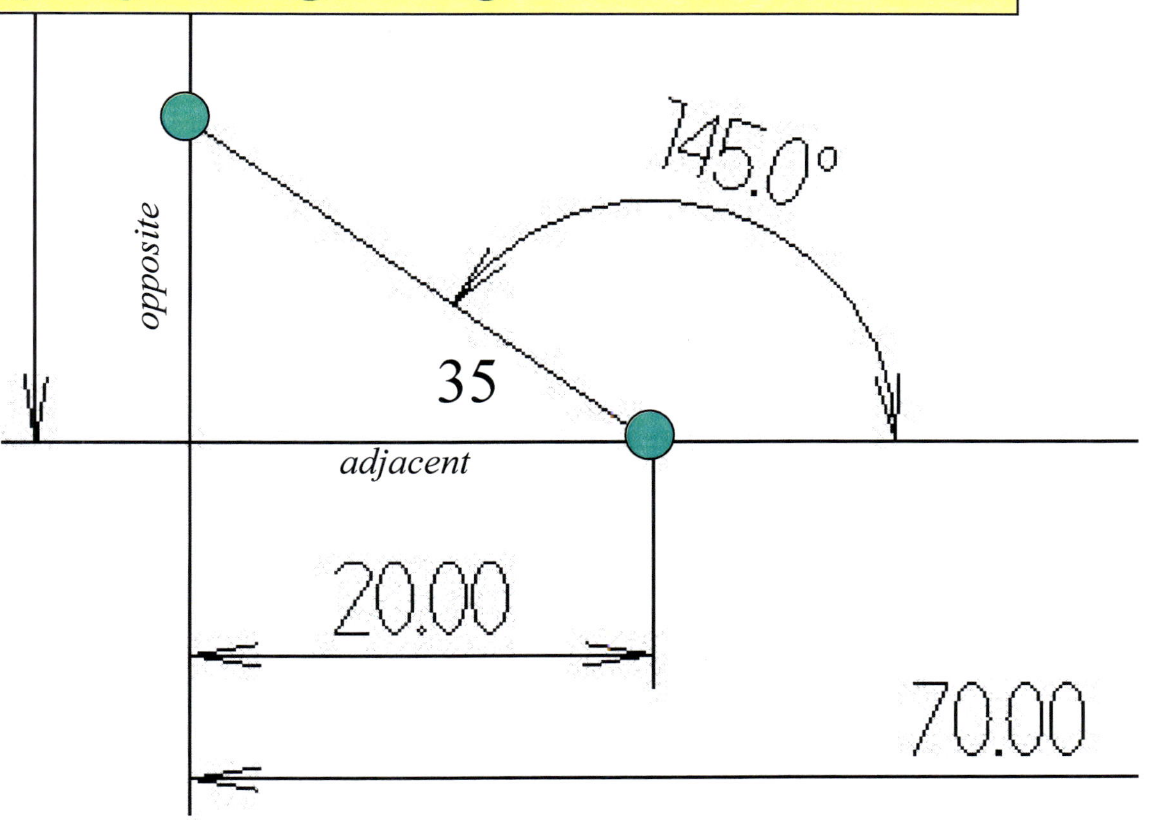

     Advanced CNC Mill Programming and Applied Mathematics Level 2

Mapping Cartesian Coordinates

After solving for the missing values, write the XY coordinates at each corner.
These coordinates become the body of your CNC program.

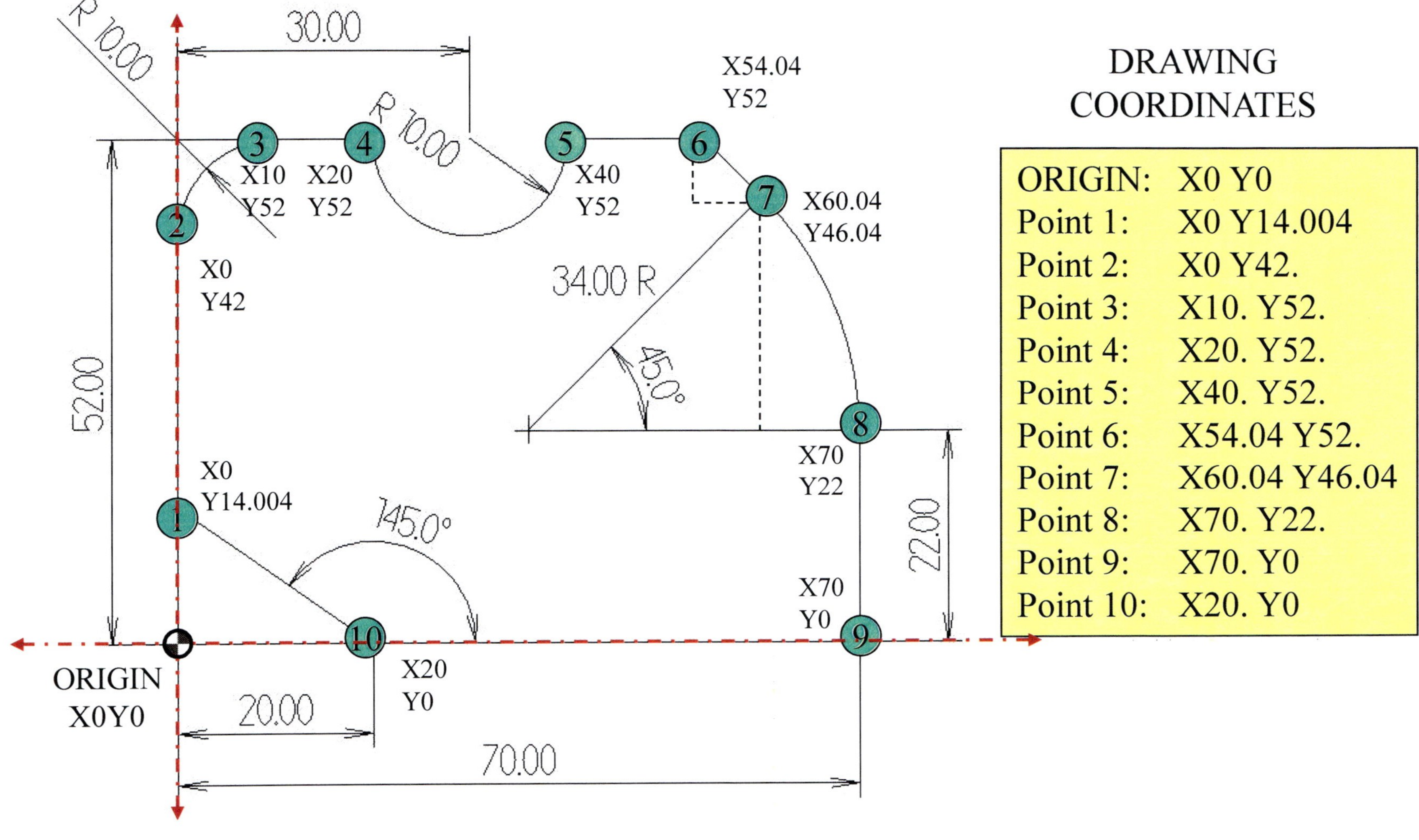

ORIGIN: X0 Y0
Point 1: X0 Y14.004
Point 2: X0 Y42.
Point 3: X10. Y52.
Point 4: X20. Y52.
Point 5: X40. Y52.
Point 6: X54.04 Y52.
Point 7: X60.04 Y46.04
Point 8: X70. Y22.
Point 9: X70. Y0
Point 10: X20. Y0

     Advanced CNC Mill Programming and Applied Mathematics Level 2

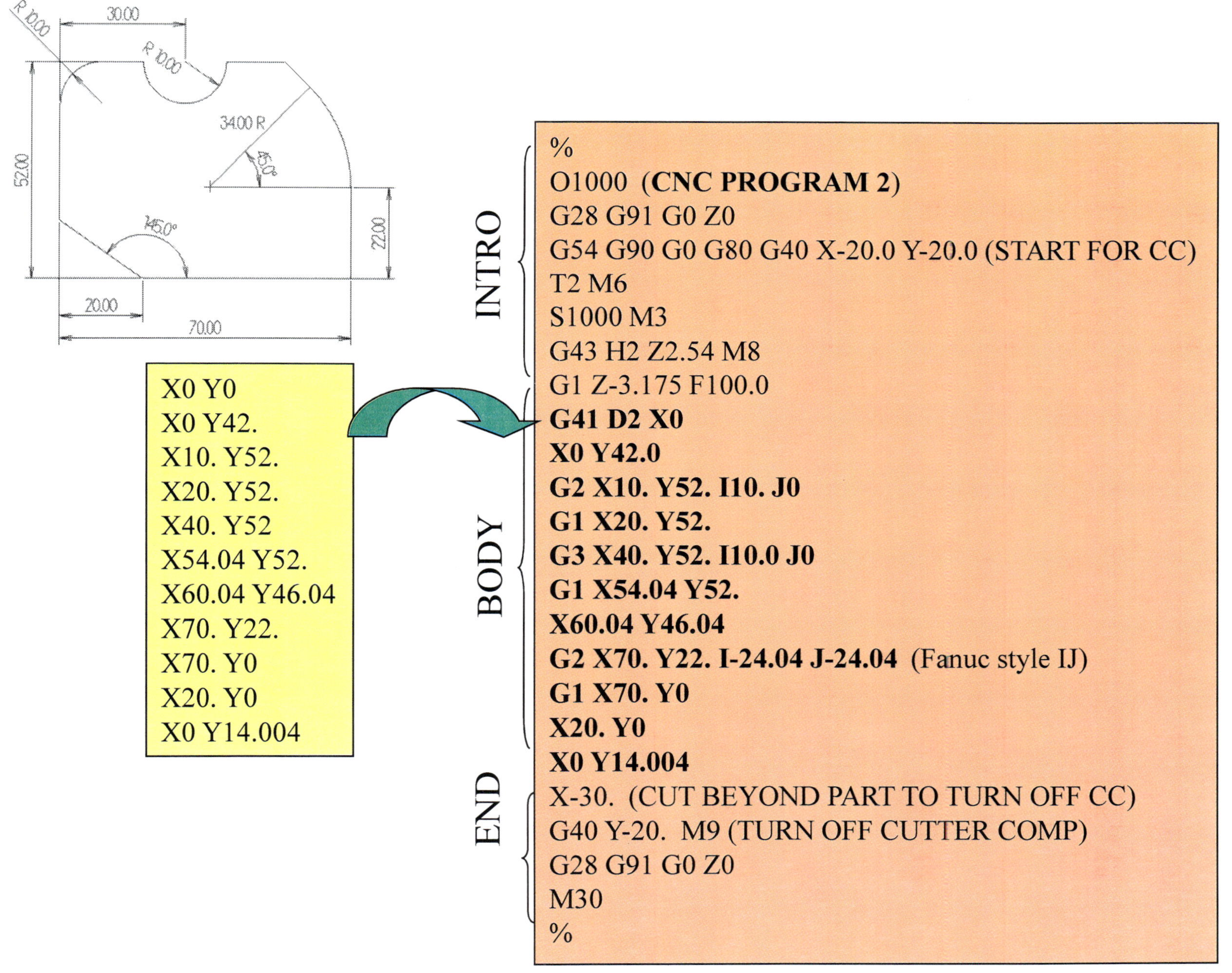

     Advanced CNC Mill Programming and Applied Mathematics Level 2

Chapter 7
Lead In & Lead Out

     Advanced CNC Mill Programming and Applied Mathematics Level 2

Objectives

1. The student will program to properly lead into the part when CNC machining to avoid dwell marks.

2. The student will program to properly lead out of the part when CNC machining to avoid dwell marks.

3. The student will demonstrate knowledge of the importance of tangency as it applies to leading in and leading out of parts.

LEAD IN/LEAD OUT

- Approaching and exiting the workpiece is important.
  - Commonly referred to as *lead-in* and *lead-out*.

- Tools should approach and exit ***tangent*** to the workpiece to avoid leaving marks.
  - The tool should always be in motion while rotating.
  - One revolution of the tool without motion is enough to leave a blemish on the workpiece.

Lead in tangentially along the edge of the workpiece.
Many times this is accomplished without using trigonometry.

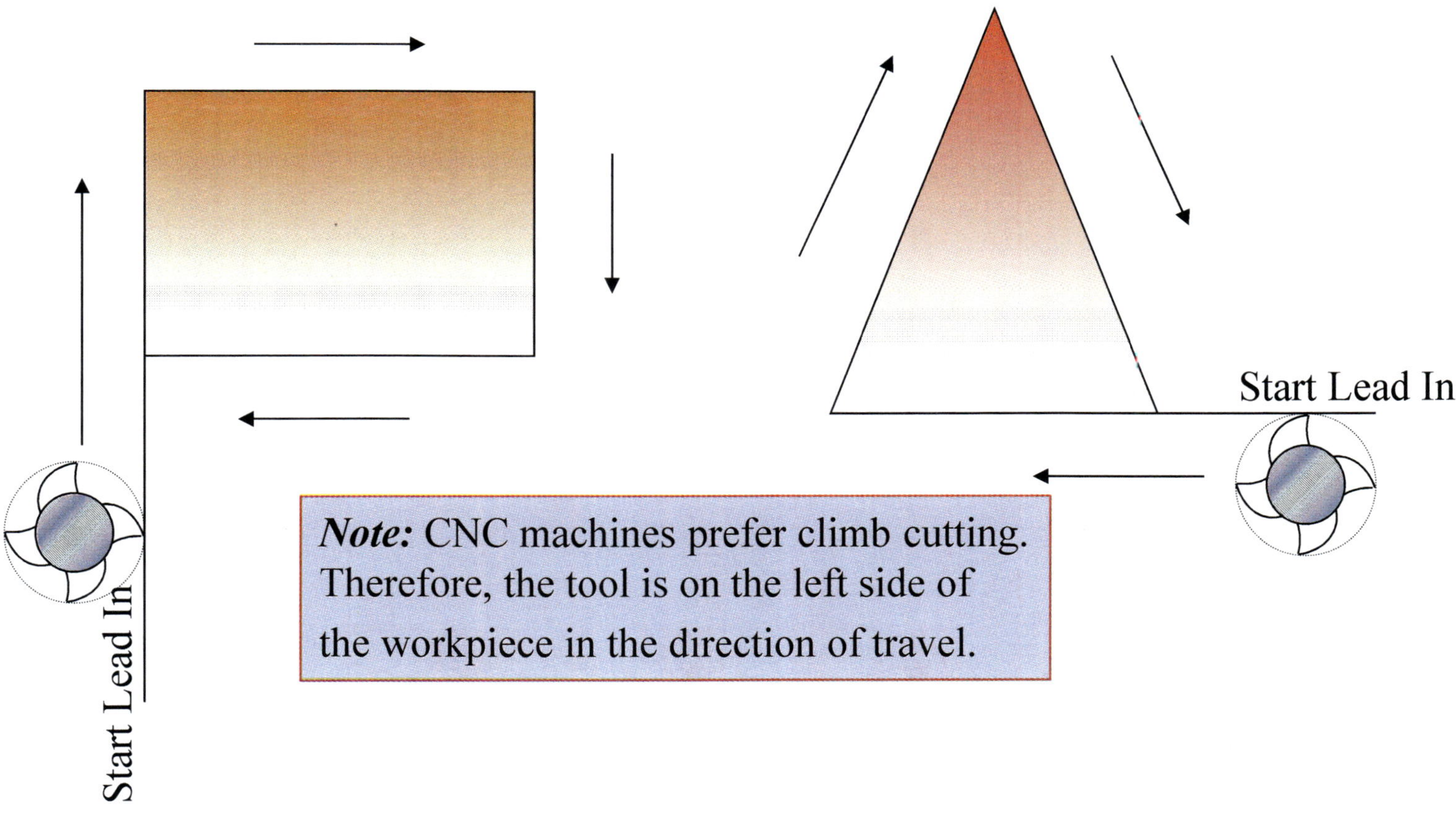

When ***exiting*** along an angled line you should extend the line tangentially instead of stopping at the theoretical tangent corner of the workpiece. You can calculate the XY coordinates for the exit point using right angle trig.

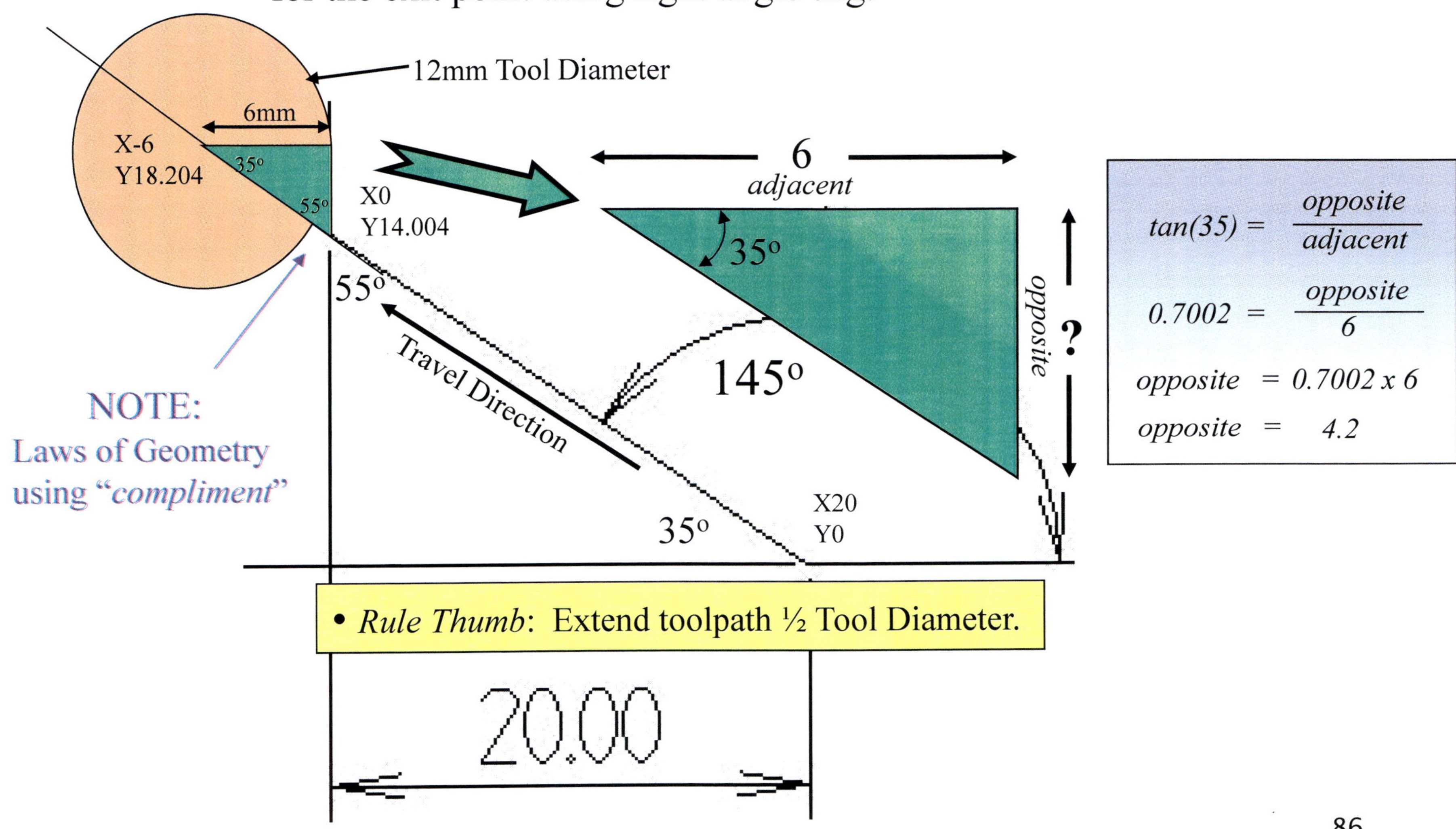

  Advanced CNC Mill Programming and Applied Mathematics Level 2

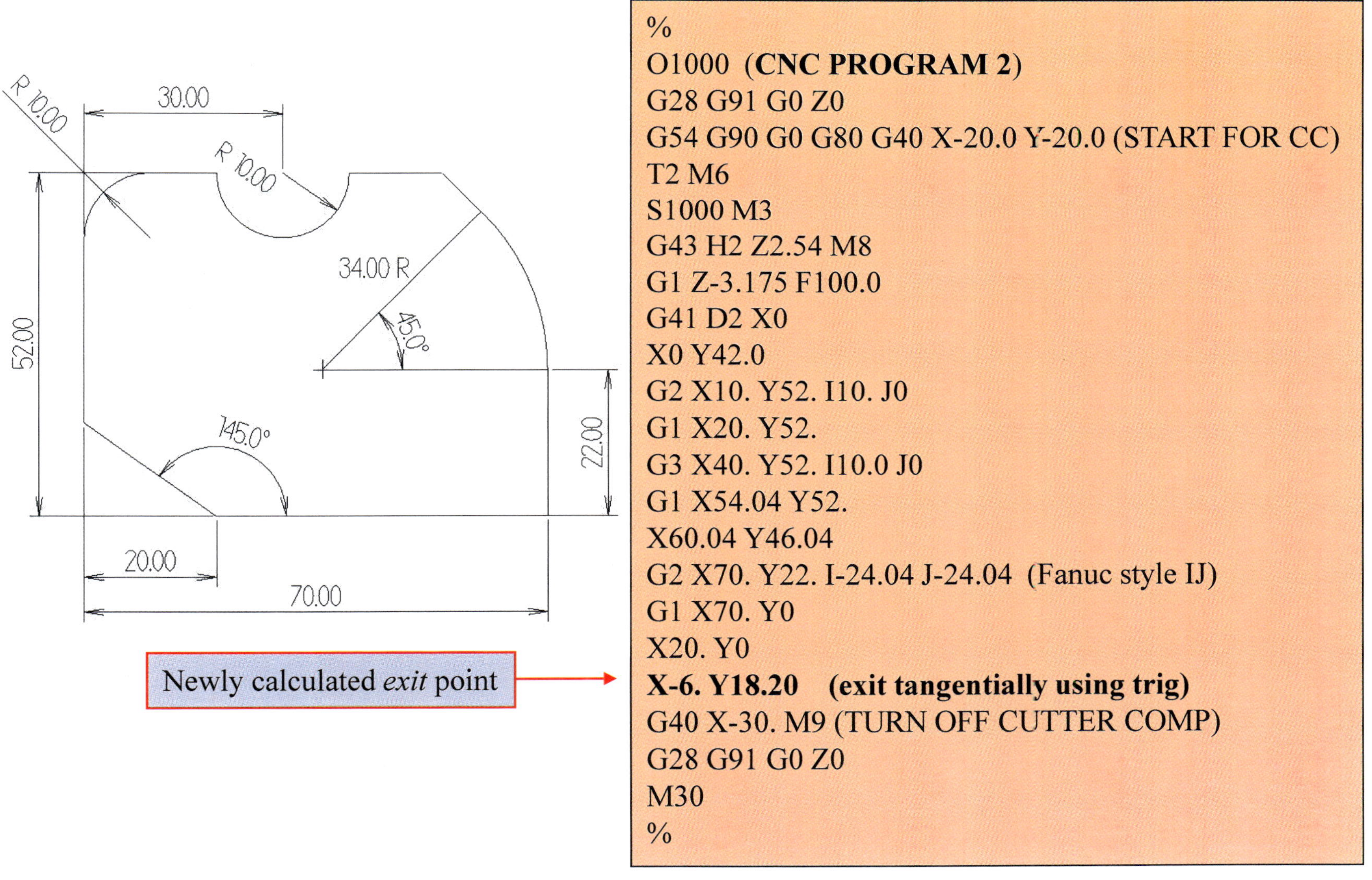


```
%
O1000  (CNC PROGRAM 2)
G28 G91 G0 Z0
G54 G90 G0 G80 G40 X-20.0 Y-20.0 (START FOR CC)
T2 M6
S1000 M3
G43 H2 Z2.54 M8
G1 Z-3.175 F100.0
G41 D2 X0
X0 Y42.0
G2 X10. Y52. I10. J0
G1 X20. Y52.
G3 X40. Y52. I10.0 J0
G1 X54.04 Y52.
X60.04 Y46.04
G2 X70. Y22. I-24.04 J-24.04  (Fanuc style IJ)
G1 X70. Y0
X20. Y0
X-6. Y18.20    (exit tangentially using trig)
G40 X-30. M9 (TURN OFF CUTTER COMP)
G28 G91 G0 Z0
M30
%
```

Leading In & Out With Arcs

When it's not feasible to lead in linearly,
lead in with an arc. The simplest method is
to use the "*baseball diamond*" approach.

1. Build a baseball diamond of a nominal size.
2. Drive the tool to first base linearly with G41.
3. Circular contour to 2nd base.
4. Machine the perimeter of the part.
5. To exit, stop at second base.
6. Circular contour to 3rd base.
7. Drive the tool home linearly with G40.

G0 X5.0 Y-1.0	(Rapid to Home Plate)
G1 G41 D1 X5.5 Y-.5	(1ST BASE; turn CC on)
G3 X5.0 Y0 R.5	(2ND BASE)

... machine part ...

G1 X5.0 Y0	(End profile cut)
G3 X-4.5 Y-.5 R.5	(3rd Base)
G1 G40 X5.0 Y-1.0	(Home Plate; turn CC off)

NOTE: When turning on cutter compensation, the minimum move must be greater than
or equal to the radius of the tool. Be sure to allow enough room should the operator switch to
another size tool.

    Advanced CNC Mill Programming and Applied Mathematics Level 2

Chapter 8
Offsetting Geometry for Centerline of Toolpath

     Advanced CNC Mill Programming and Applied Mathematics Level 2

Objectives

1. The student will offset part geometry by a distance of one half the tool diameter using the principles of geometry.

2. The student calculate tool center points using the principles of right angle trig.

Offsetting Geometry to Define Center of Tool Path
Tool Diameter = 1.0

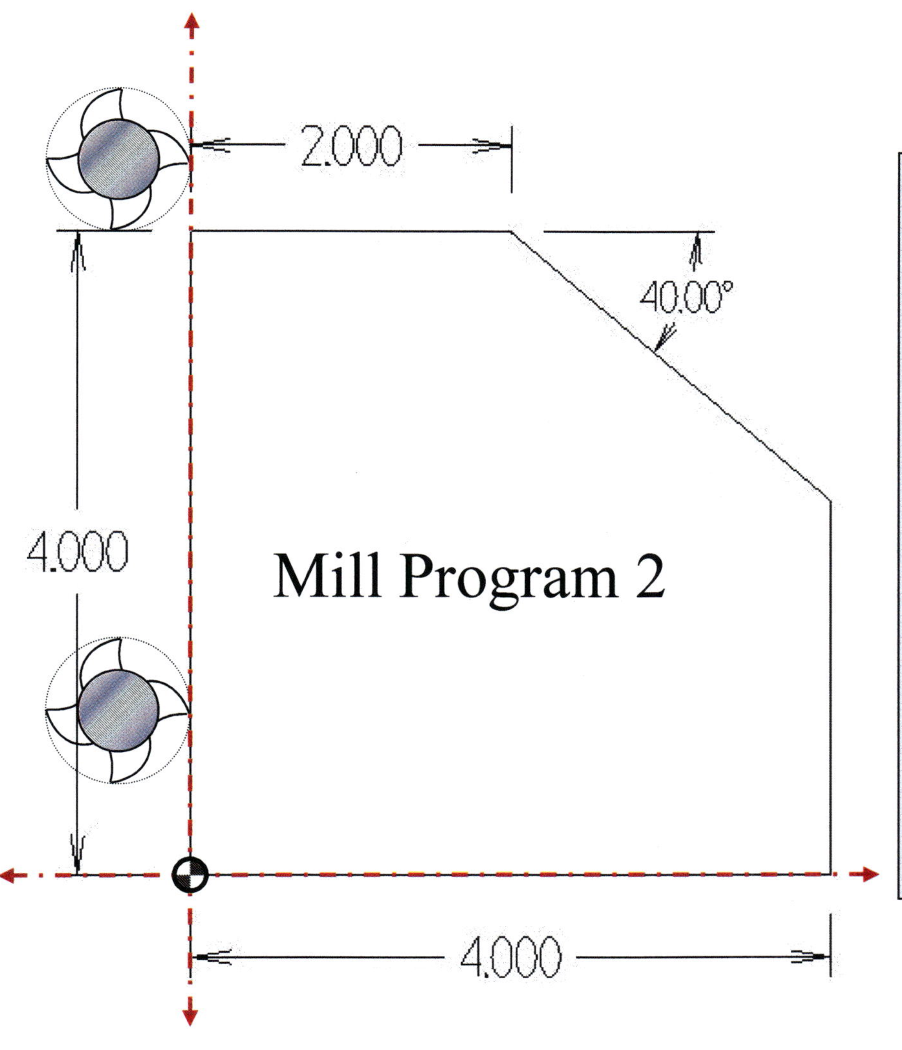

Offsetting geometry for center of toolpath is necessary when you don't use Cutter Compensation (CC).

Offsetting for vertical and horizontal walls is easy because you simply add or subtract the tool radius perpendicular to the axis.

To offset along *angled* walls, however, **geometry** must be used to determine the angle of offset, followed by **right angle trig** to calculate the lengths of the legs along the XY axis (mill) or ZX axis (lathe).

For this reason, it's recommended to use CC whenever possible save time and to reduce errors.

Offsetting Geometry for Center of Tool Path

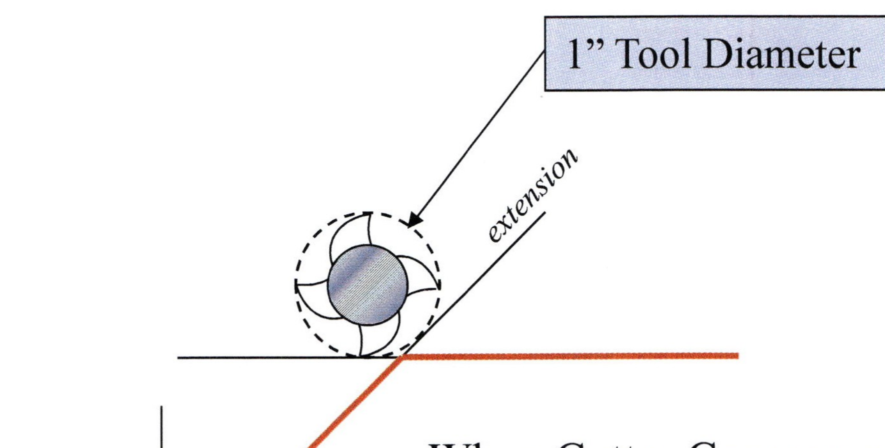
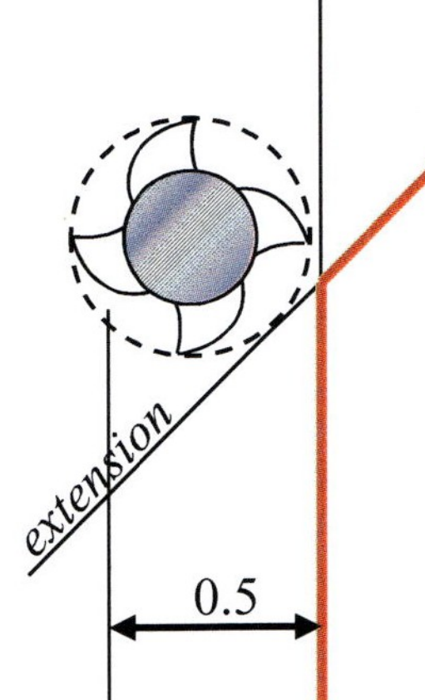

When Cutter Compensation is not available, the toolpath must be offset by an amount equal to the tool radius. This poses two problems:
1. Geometry and right angle trig must be used to calculate both end points.
2. The Operator must use the exact diameter tool called out in the program—with no option for substitutes.

     Advanced CNC Mill Programming and Applied Mathematics Level 2

Recipe For Offsetting Geometry

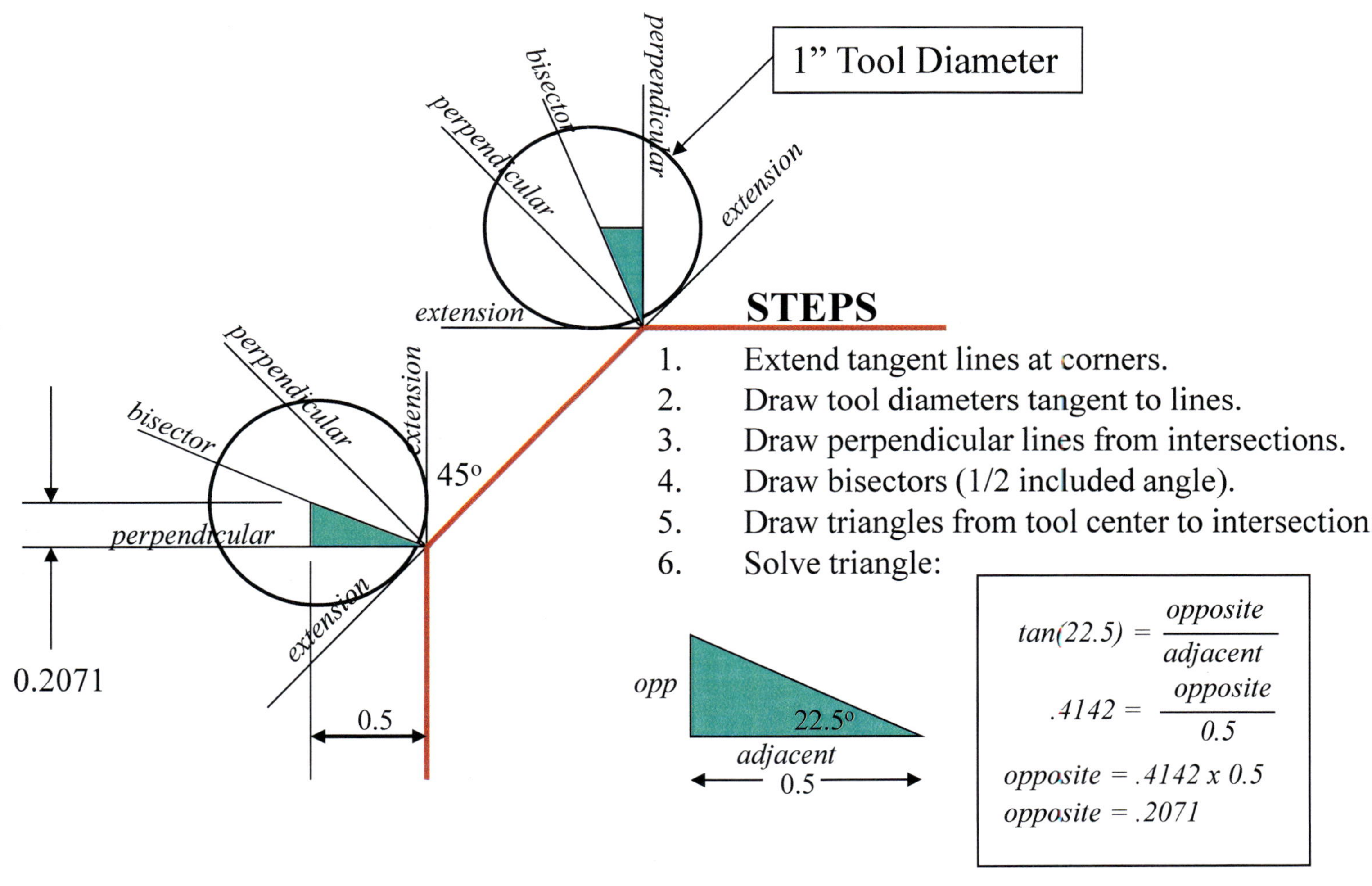

$$tan(22.5) = \frac{opposite}{adjacent}$$

$$.4142 = \frac{opposite}{0.5}$$

$$opposite = .4142 \times 0.5$$

$$opposite = .2071$$

    Advanced CNC Mill Programming and Applied Mathematics Level 2

CNC Offset Program 3

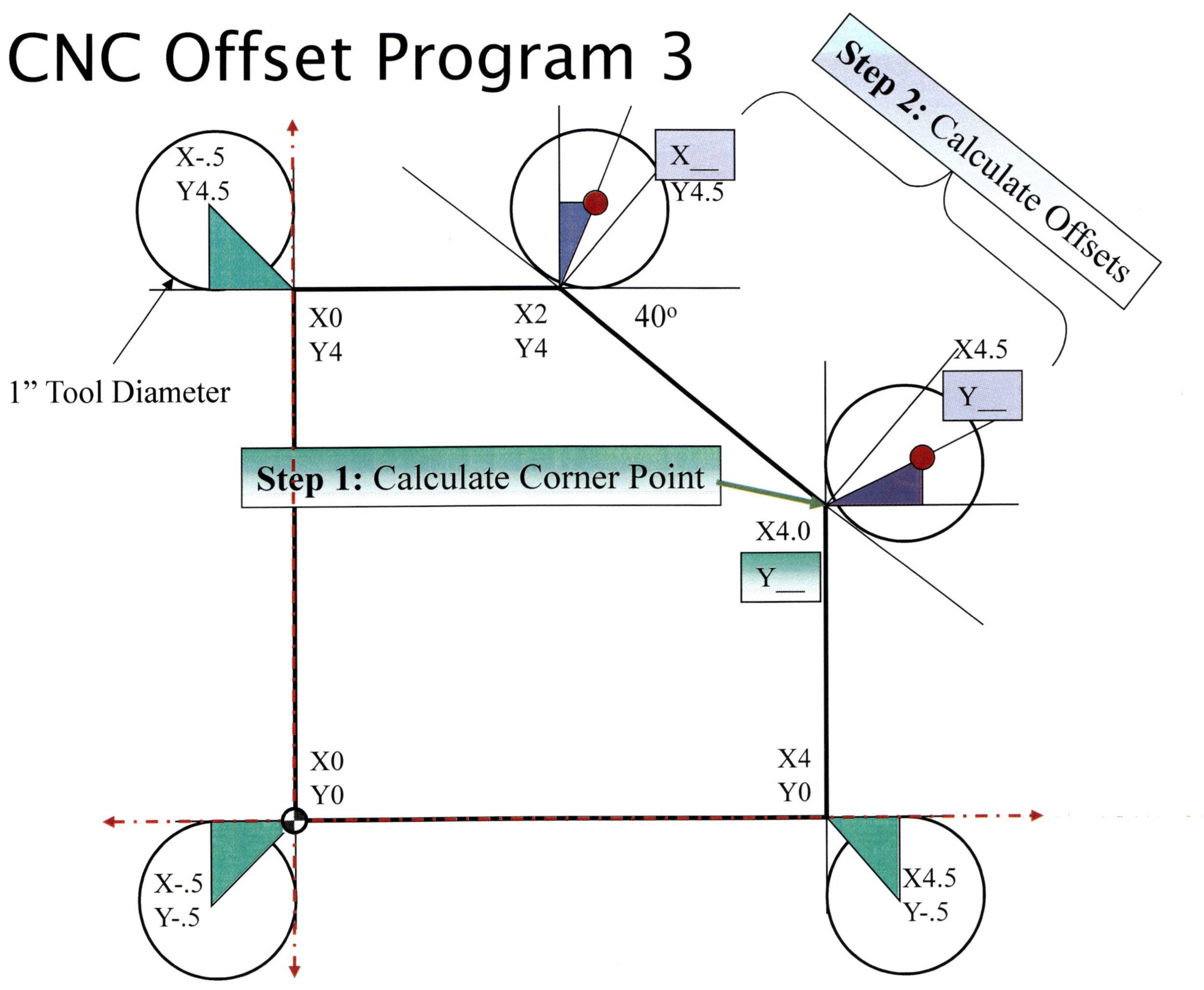

     Advanced CNC Mill Programming and Applied Mathematics Level 2

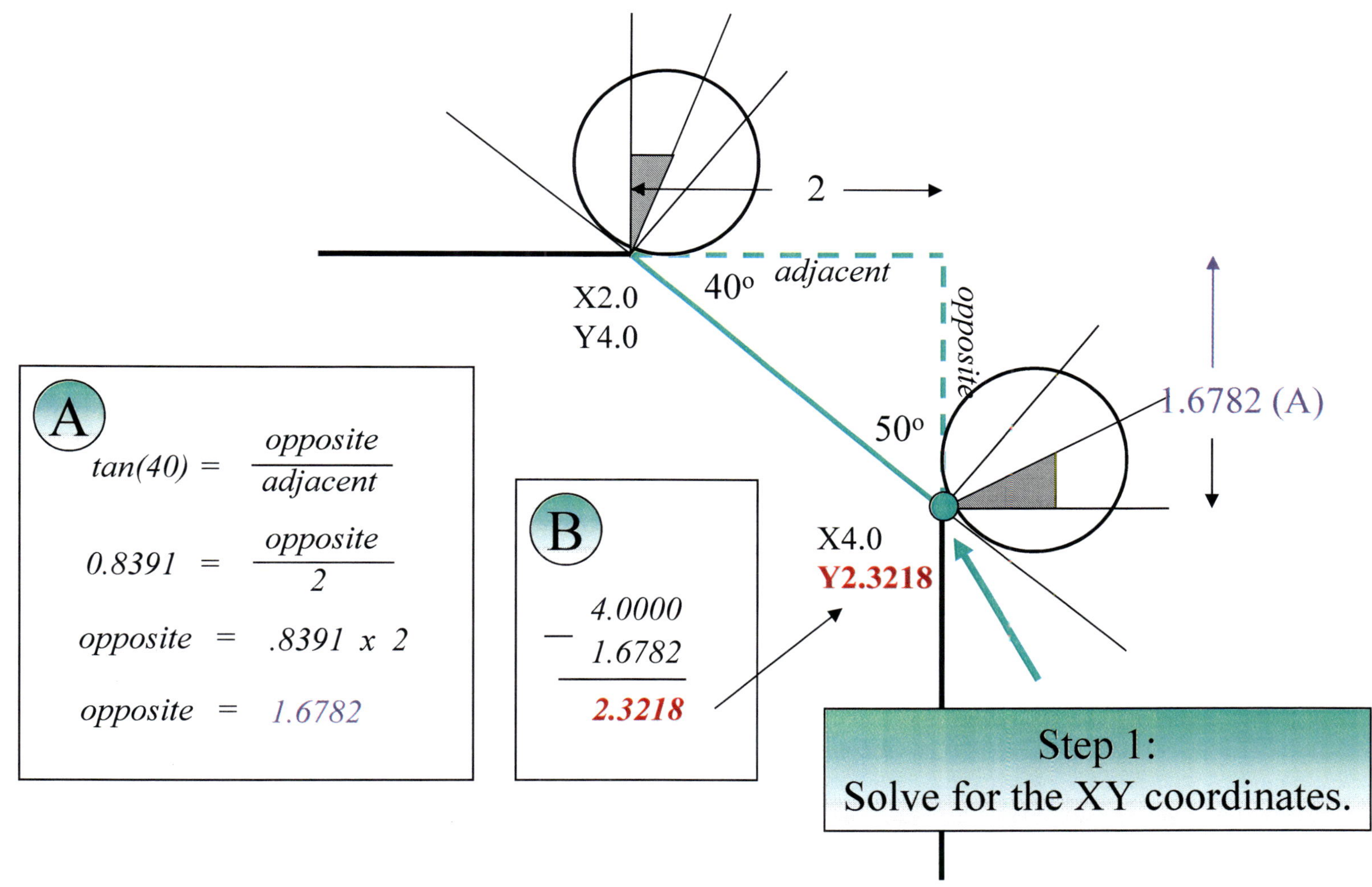

2
X2.0
Y4.0
40º
adjacent
opposite
50º
1.6782 (A)
A
tan(40) = opposite/adjacent
0.8391 = opposite/2
opposite = .8391 x 2
opposite = 1.6782
B
X4.0
Y2.3218
4.0000
− 1.6782
2.3218
Step 1:
Solve for the XY coordinates.

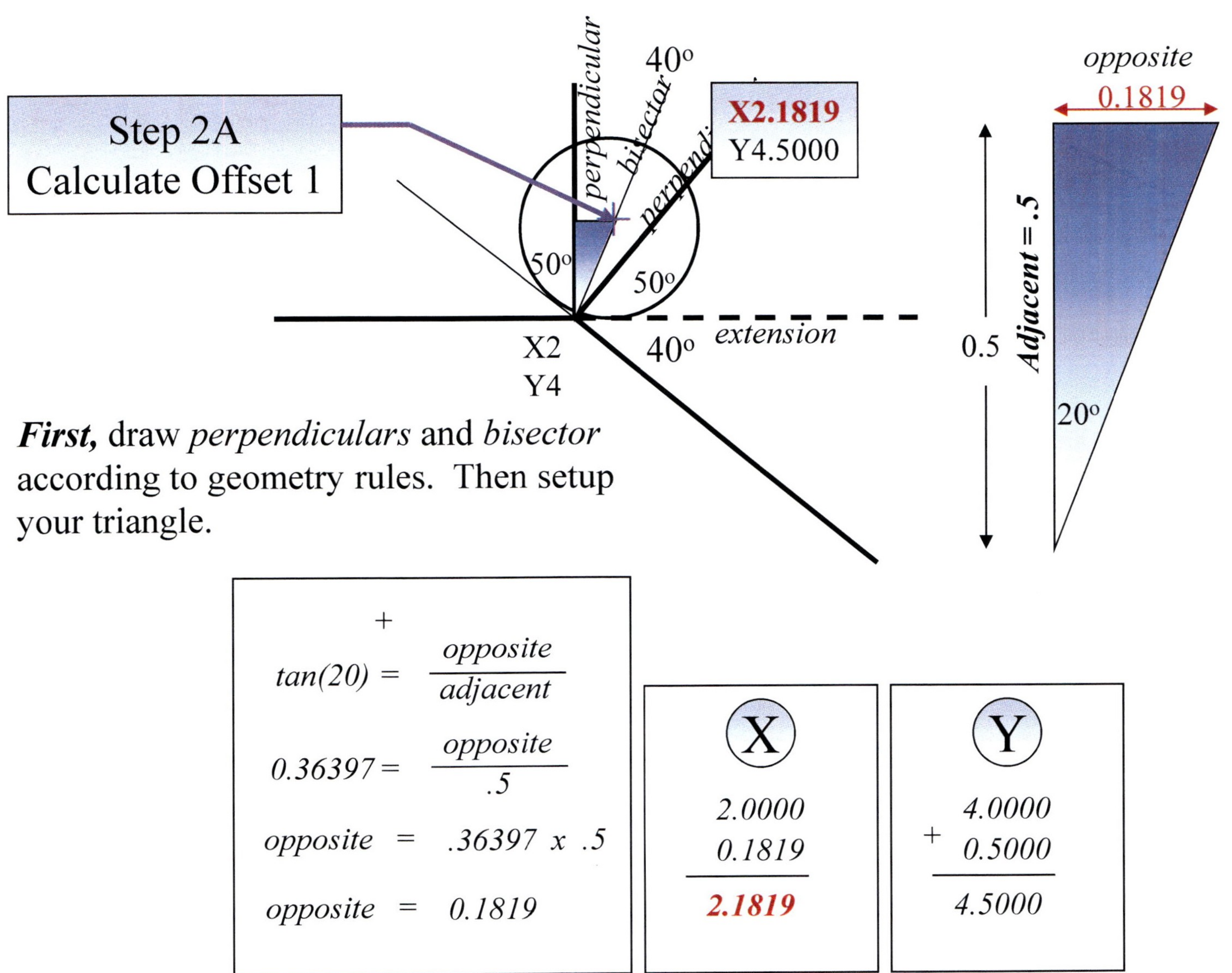

First, draw *perpendiculars* and *bisector* according to geometry rules. Then setup your triangle.

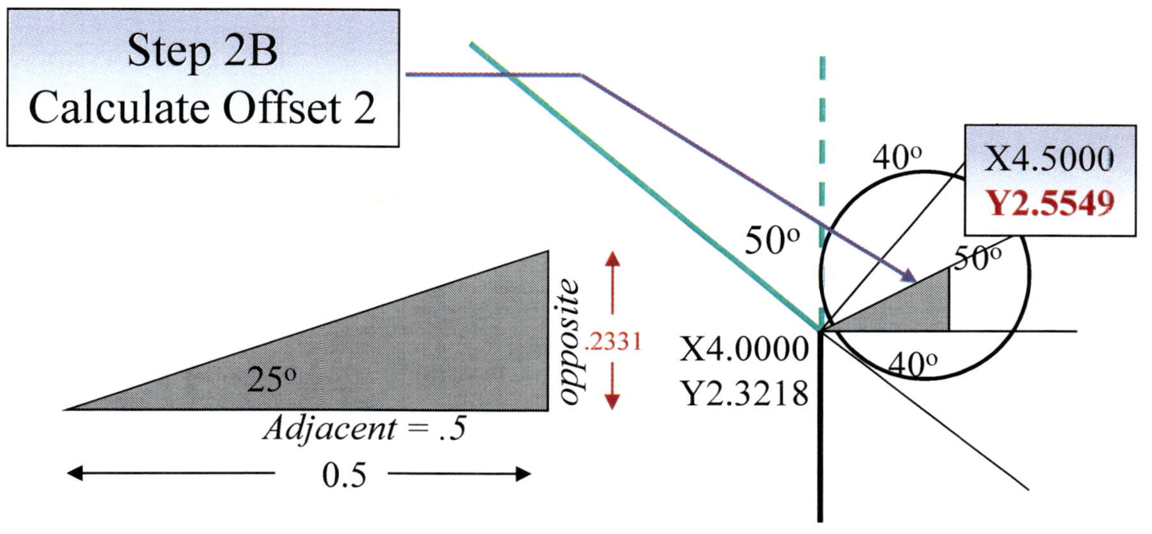

$$tan(25) = \frac{opposite}{adjacent}$$

$$0.4663 = \frac{opposite}{.5}$$

$$opposite = .4663 \times .5$$

$$opposite = 0.2331$$

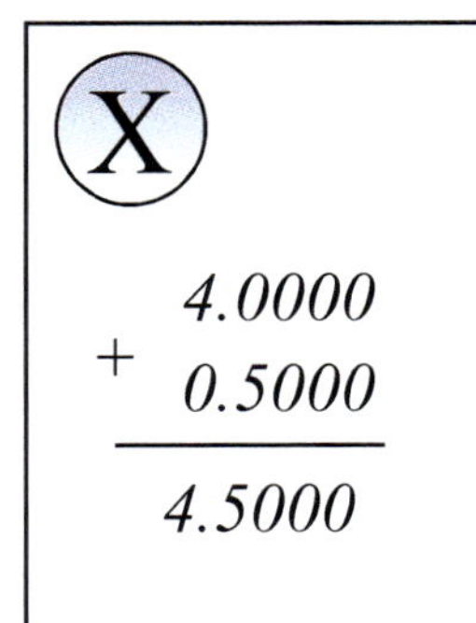

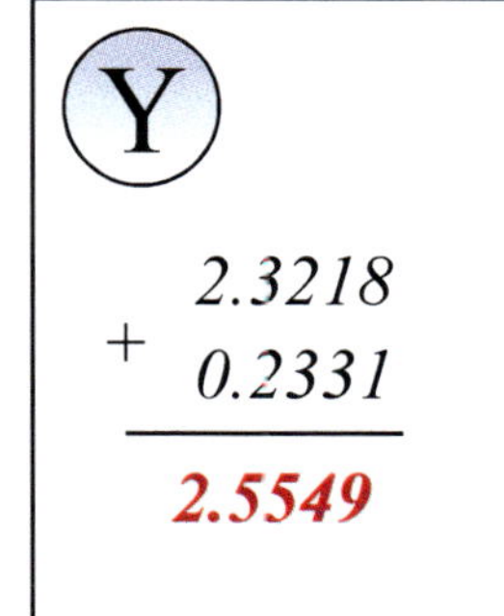

Mapping Cartesian Coordinates

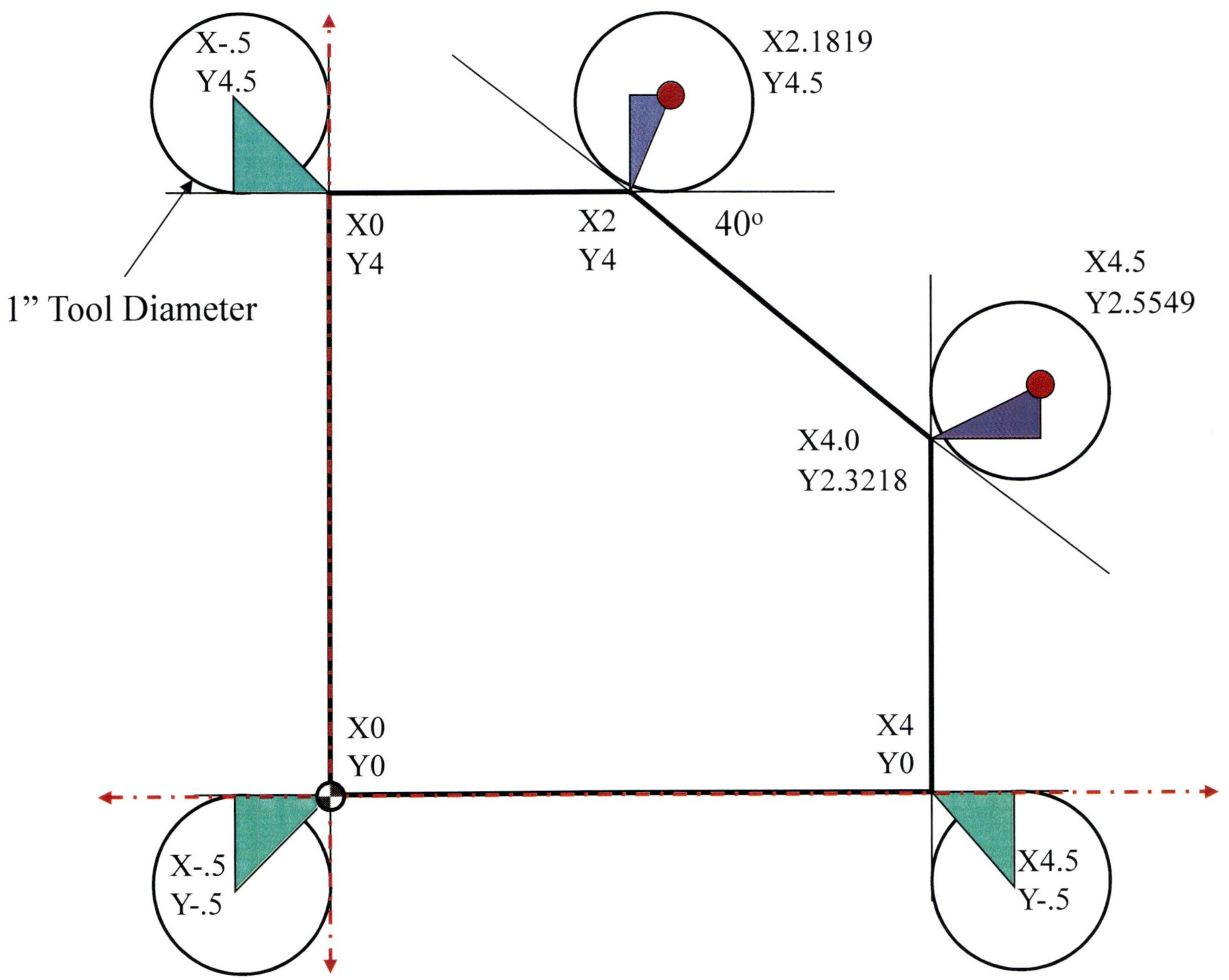

     Advanced CNC Mill Programming and Applied Mathematics Level 2

Linear Interpolation Offset CNC Program 3

WITH CC - No Offset

```
O1000  (MILL EXERCISE 3)
G28 G91 G0 Z0
G54 G90 G0 X-1.0 Y-1.0
T2 M6
S1000 M3
G43 H2 Z0.1 M8
G1 Z-0.1 F10.0
G41 D2 X0
X0 Y4.0
G1 X2.0 Y4.0
G1 X4.0 Y2.3218 (trig corner)
X4.0 Y0
X-1.0
G40 Y-1.0 (turn off CC)
G28 G91 G0 Z0
M30
```

WITHOUT CC - OFFSET

```
O1000  (MILL EXERCISE 3)
G28 G91 G0 Z0
G54 G90 G0 X-0.5 Y-0.5
T2 M6
S1000 M3
G43 H2 Z0.1 M8
G1 Z-0.1 F10.0
X-0.5
X-0.5 Y4.5
G1 X2.1819 Y4.5 (trig offset X)
G1 X4.5 Y2.5549 (trig offset Y)
X4.5 Y-0.5
X-0.5

G28 G91 G0 Z0
M30
```

APPENDIX A
CNC Mill Programming
Level 2 Problems & Solutions

Mill Project 301; Drill 12 Holes

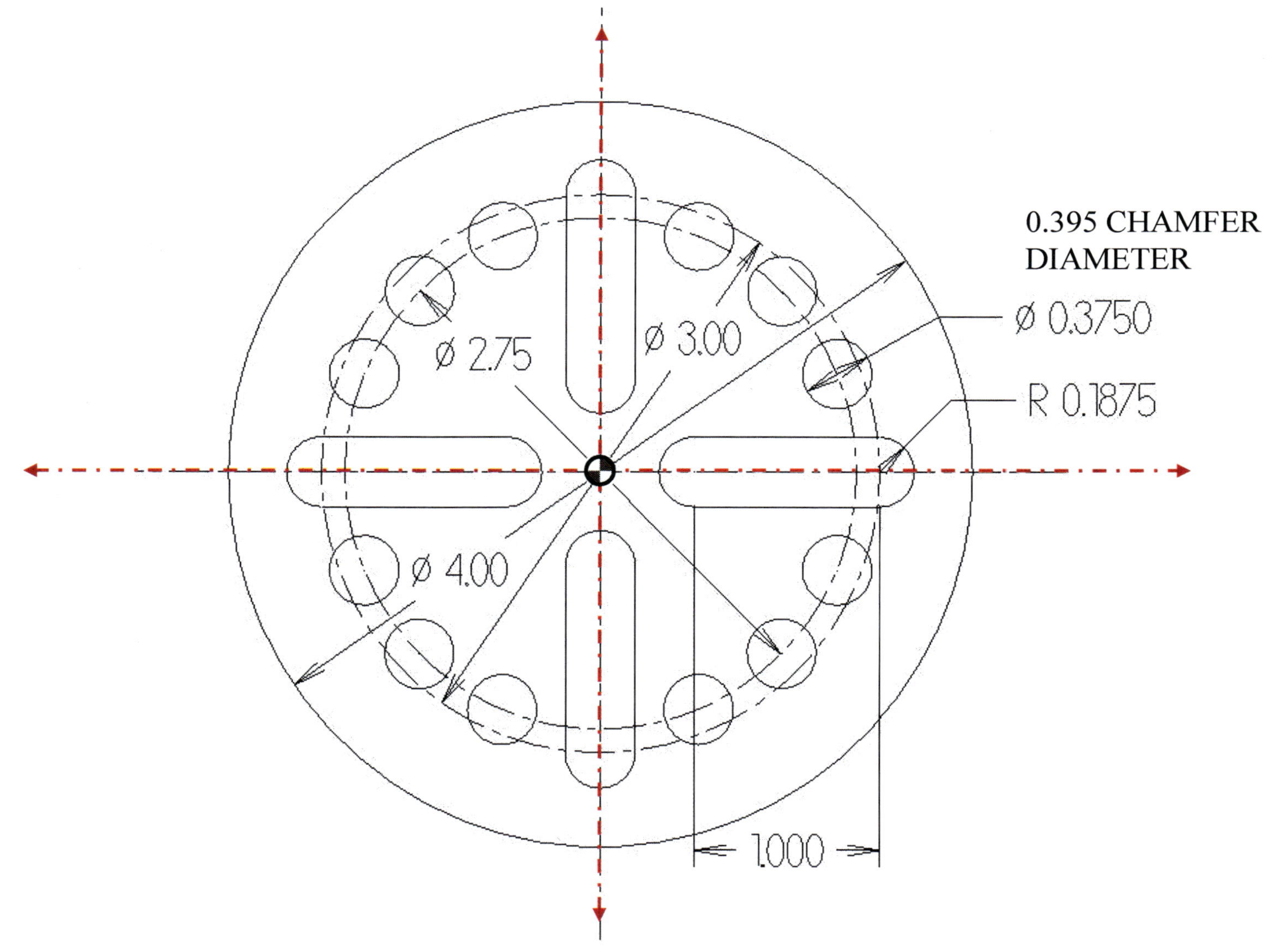

Mill Project 301

Trig Functions:

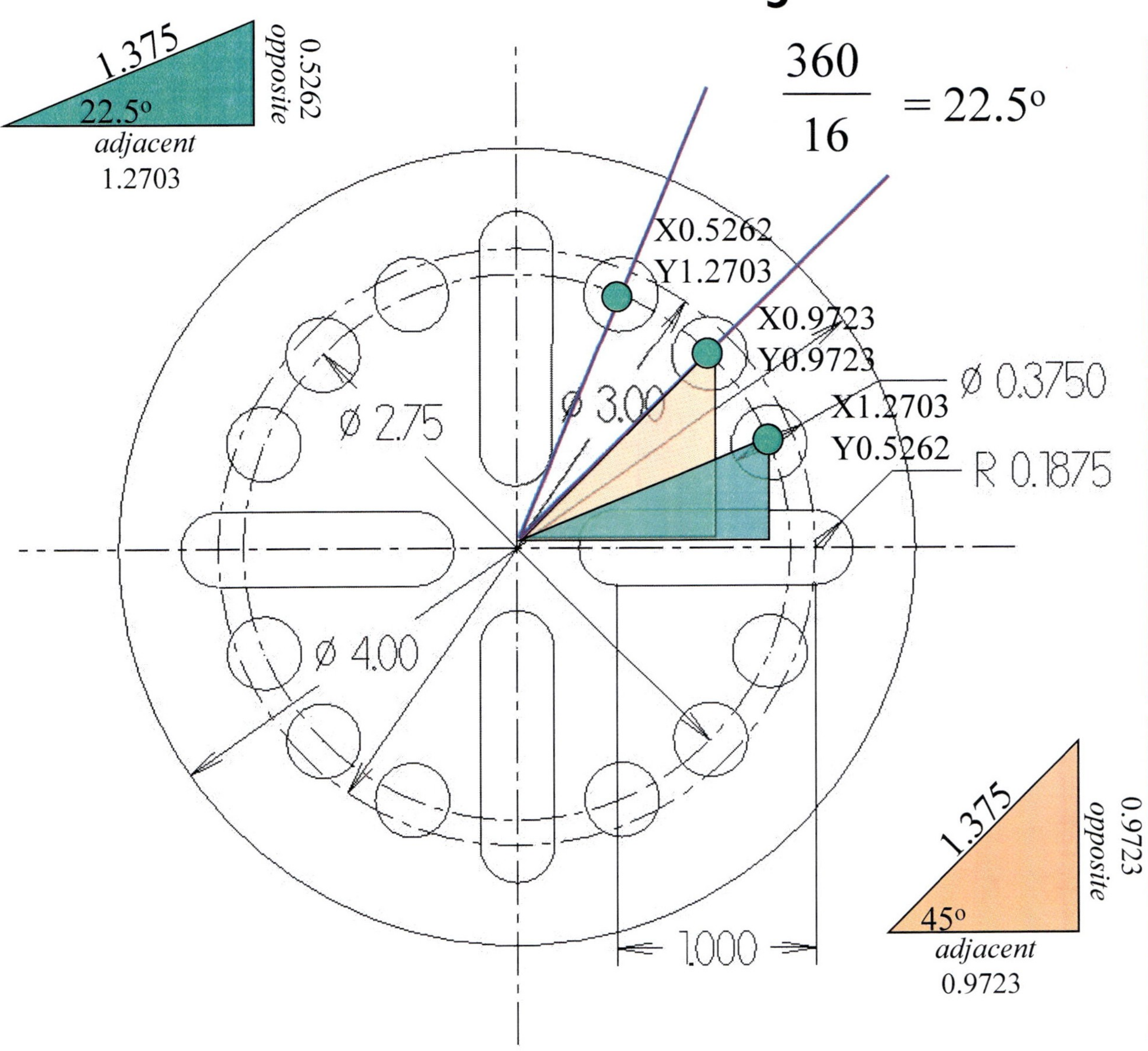

$$\frac{360}{16} = 22.5°$$

$$sin(22.5) = \frac{opposite}{hypotenuse}$$

$$sin(22.5) = \frac{opposite}{1.375}$$

$$0.3827 = \frac{opposite}{1.375}$$

$$opposite = 0.3827 \times 1.375$$

$$opposite = 0.5262$$

$$cos(22.5) = \frac{adjacent}{hypotenuse}$$

$$cos(22.5) = \frac{adjacent}{1.375}$$

$$0.9239 = \frac{adjacent}{1.375}$$

$$adjacent = 0.9239 \times 1.375$$

$$adjacent = 1.2703$$

$$sin(45) = \frac{opposite}{hypotenuse}$$

$$sin(45) = \frac{opposite}{1.375}$$

$$0.7071 = \frac{opposite}{1.375}$$

$$opposite = 0.7071 \times 1.375$$

$$opposite = 0.9723$$

   Advanced CNC Mill Programming and Applied Mathematics Level 2

Polar Coordinates (*R theta*)

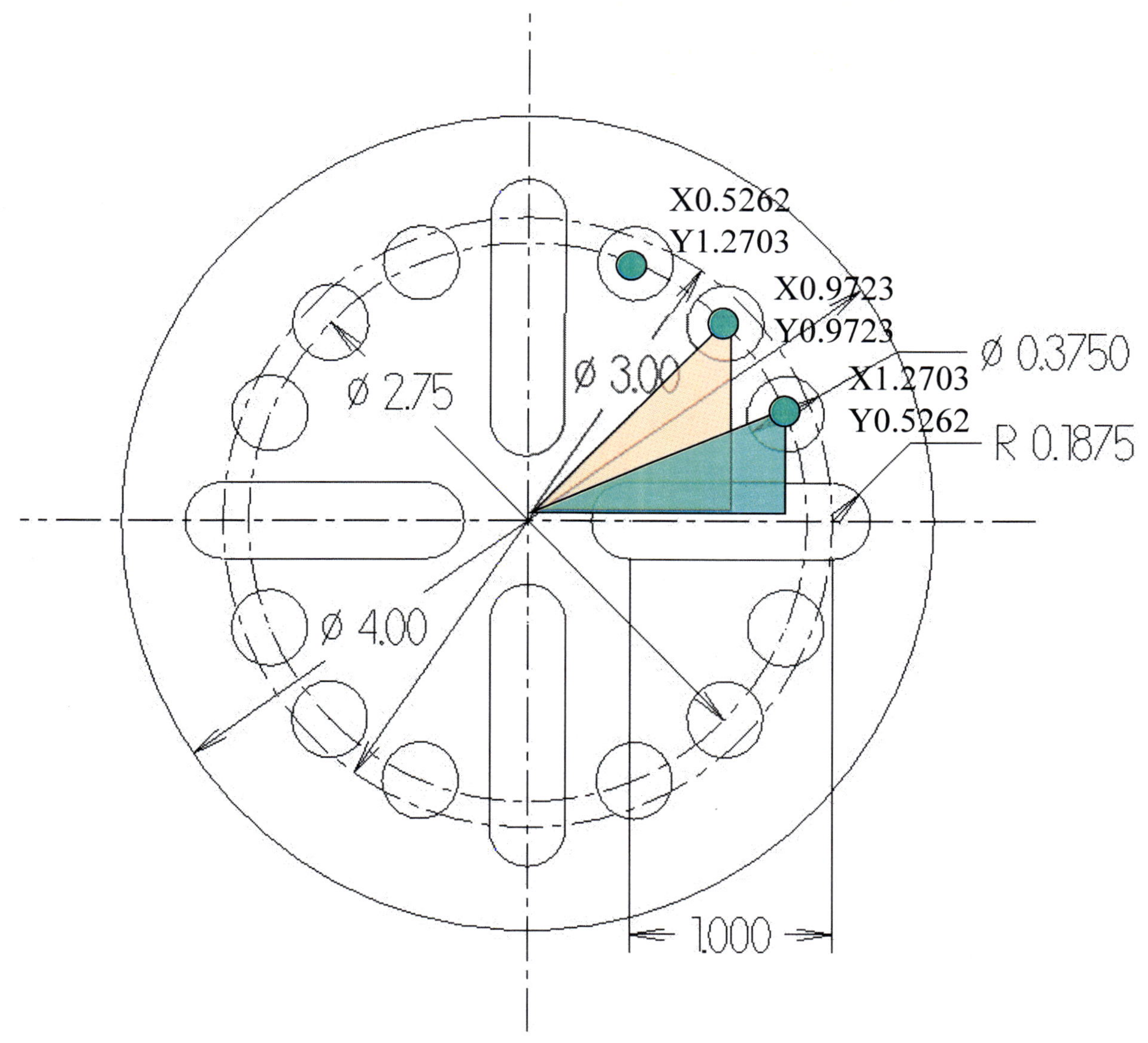

Y = r x sin(22.5)

Y = 1.375 x 0.38268

Y = 0.5262

X = r x cos(22.5)

X = 1.375 x 0.9239

X = 1.2703

X = r x cos(45)

X = 1.375 x 0.7071

X = 0.9723

Mill Project 301 PROGRAM

```
%
O301 ( ORIGIN CENTER )
G54 G90 G0 X0 Y0
T3 M6  (90 DEG SPOT DRILL)
S1000 M3
G43 H3 Z0.1 M8

( ## QUADRANT 1 ## )

G81 X1.2703 Y0.5262 Z-0.1975 R0.1 F10.
X0.9723 Y0.9723
X0.5262 Y1.2703


( ## QUADRANT 2 ## )
X-0.5262 Y1.2703
X-0.9723 Y0.9723
X-1.2703 Y0.5262
```

```
( ## QUADRANT 3 ## )
X-1.2703 Y-0.5262
X-0.9723 Y-0.9723
X-0.5262 Y-1.2703


( ## QUADRANT 4 ## )
X0.5262 Y-1.2703
X0.9723 Y-0.9723
X1.2703 Y-0.5262
G80 M9


G28 G91 G0 Z0
M30
%
```


   Advanced CNC Mill Programming and Applied Mathematics Level 2

Mill Project 302

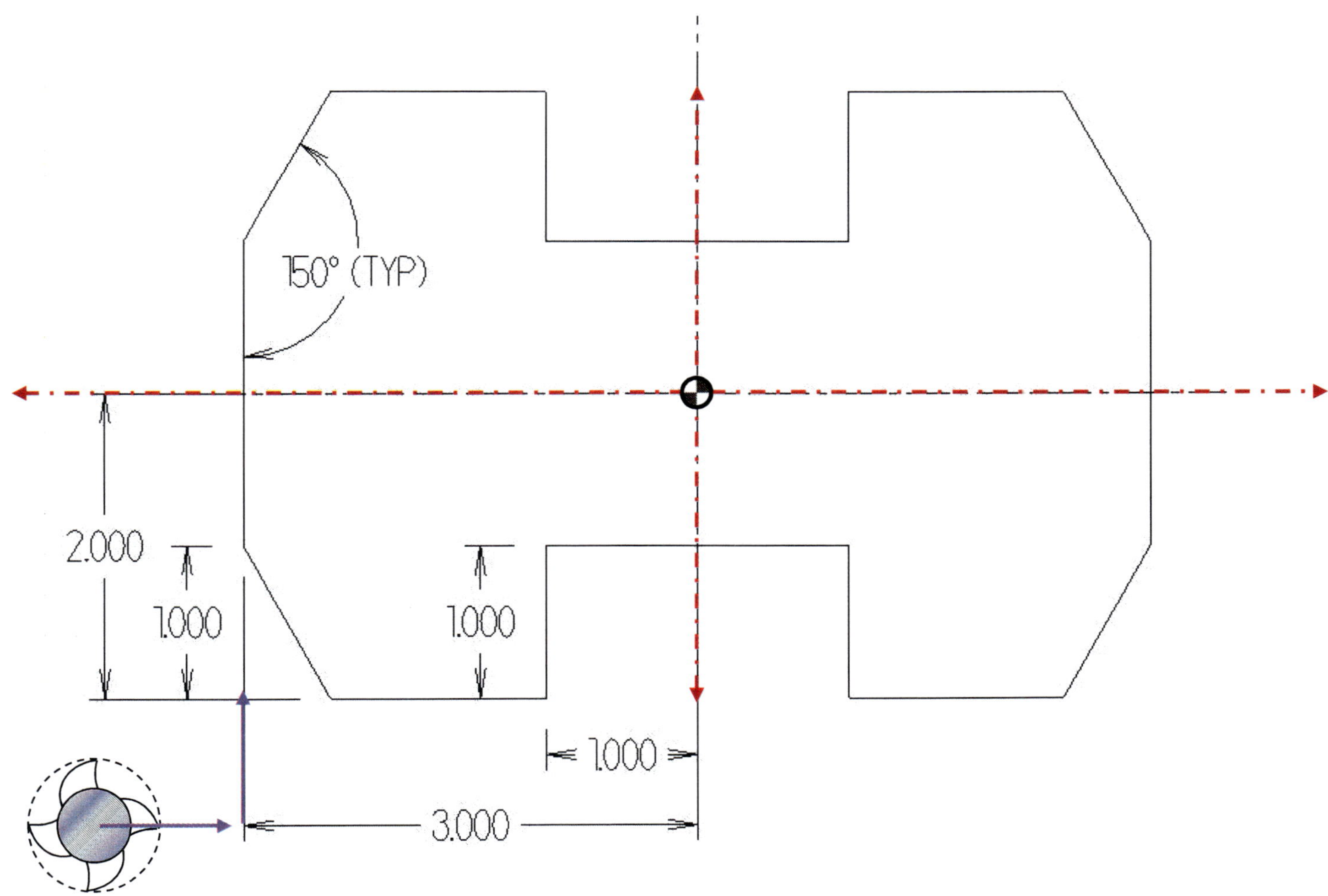

Mill Project 302

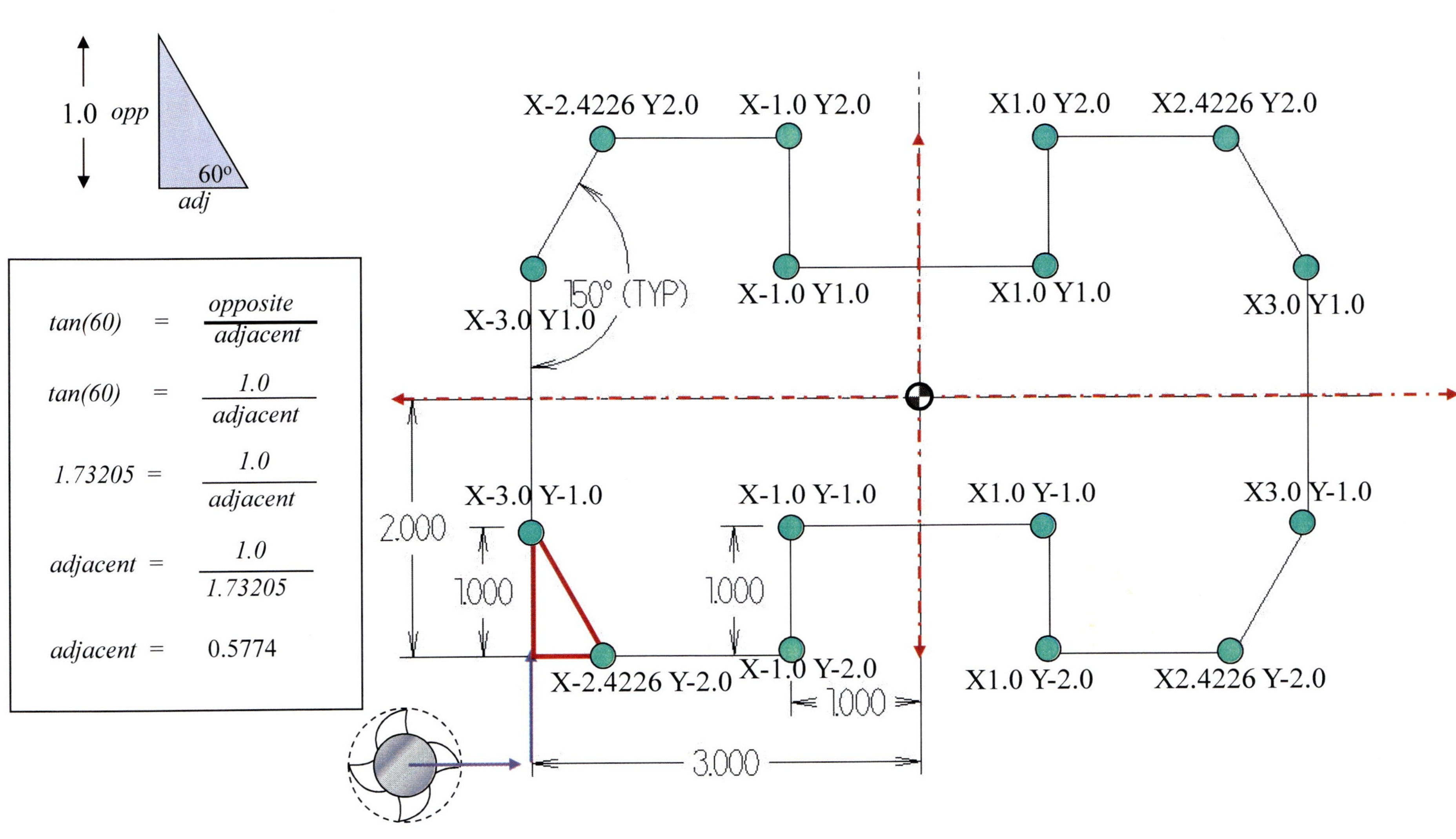

   Advanced CNC Mill Programming and Applied Mathematics Level 2

Mill Project 302 PROGRAM

```
%
O302 ( ORIGIN CENTER )
      ( TOOL ON )
G54 G90 G0 X-4.0 Y-3.0 G80 G40
T2 M6
S1000 M3
G43 Z0.1
G1 H2 Z-0.1 F10. M8
G41 D2 X-3.0
Y1.0
X-2.4226 Y2.0 (CHAMFER 1)
X-1.0
Y1.0
X1.0
Y2.0
X2.4226
X3.0 Y1.0
Y-1.0
X2.4226 Y-2.0
G1 X1.0
Y-1.0
X-1.0
Y-2.0
X-2.4226
X-3.0 Y-1.0
X-4.0
G40 Y-3.0
G28 G91 G0 Z0
M30
%
```


     Advanced CNC Mill Programming and Applied Mathematics Level 2

Mill Project 303

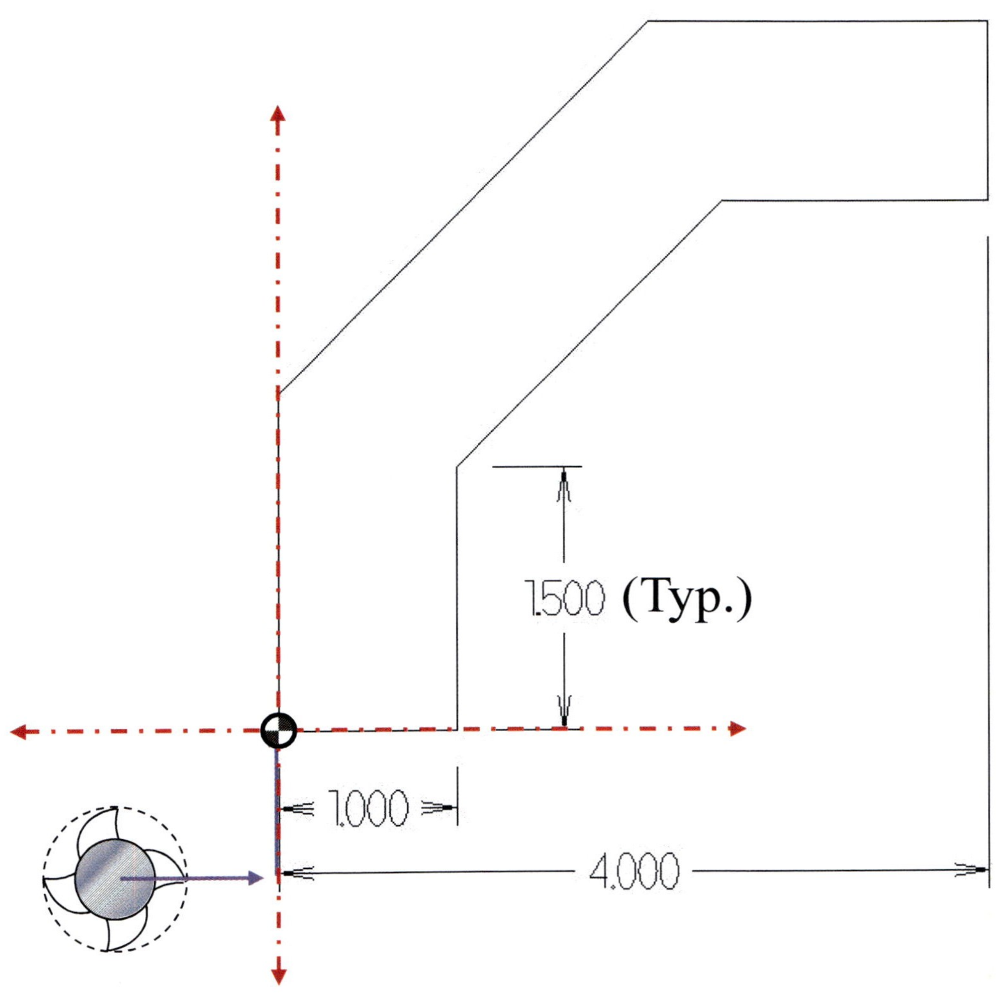

Advanced CNC Mill Programming and Applied Mathematics Level 2

Mill Project 303

Step 1: Draw perpendicular lines.

Step 2: Draw bisectors.

Step 3: Draw triangle legs.

Step 4:

Determine known values:

  Inside Angle   = 22.5

  Side Adjacent =   1.0

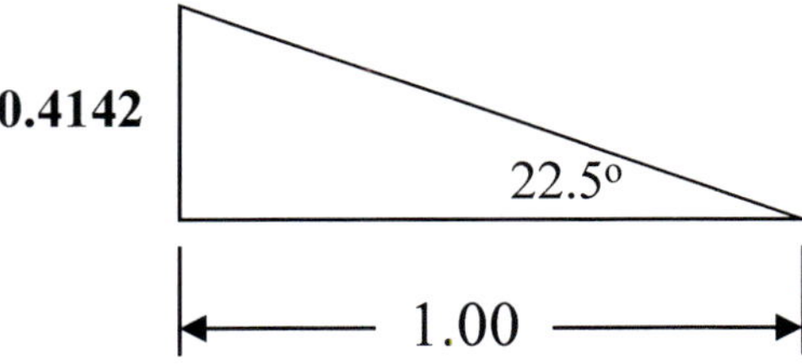

Step 5:

Solve for Side Opposite:

$$tan(22.5) = \frac{opposite}{adjacent}$$

$$tan(22.5) = \frac{opposite}{1.0}$$

$$0.4142 = \frac{opposite}{1.0}$$

$$opposite = 0.4142 \times 1.0$$

$$opposite = \mathbf{0.4142}$$

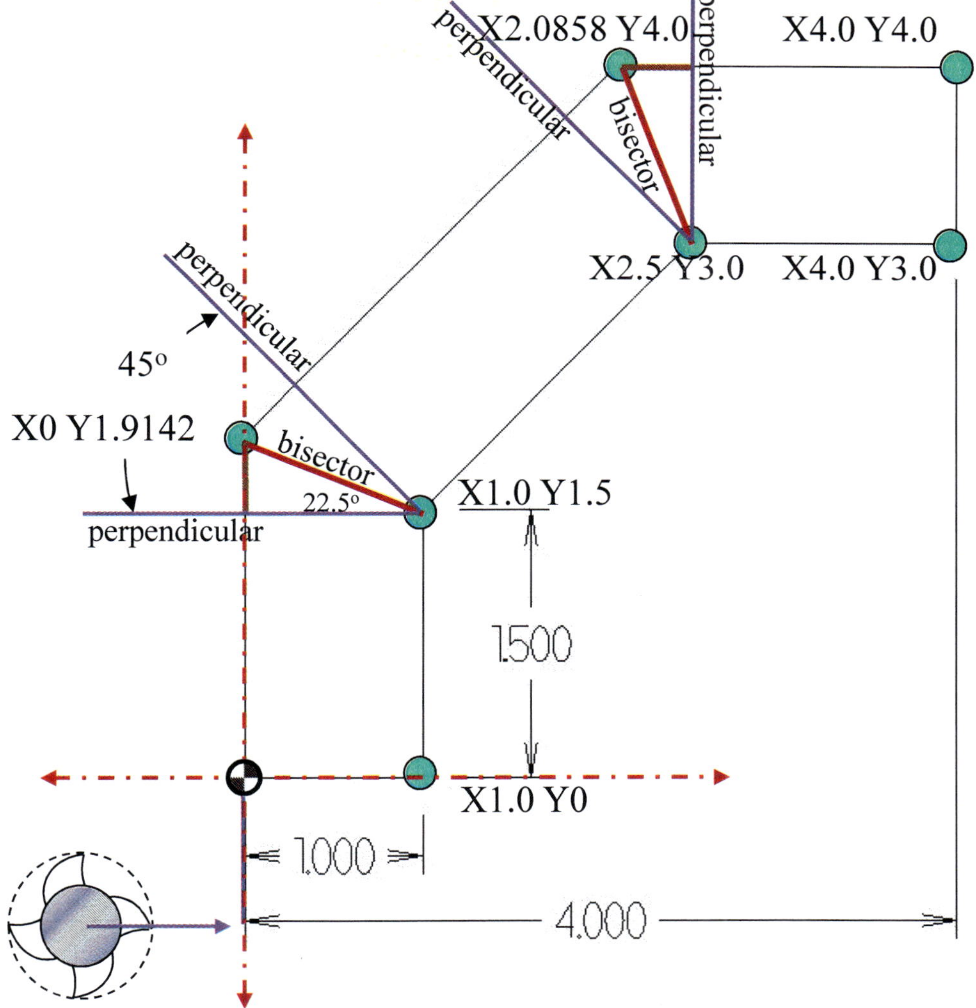

Mill Project 303 PROGRAM

```
%
O303 ( ORIGIN LOWER-LEFT )
     ( TOOL ON )
G54 G90 G0 X-1.0 Y-1.0
T2 M6
S1000 M3
G43 H2 Z0.1
G1 Z-0.1 F10.0 M8
G41 D2 X0
Y1.9142
X2.0858 Y4.0
X4.0
Y3.0
X2.5
X1.0 Y1.5
Y0
X-1.0
G40 Y-1.0
G28 G91 G0 Z0
M30
%
```


   Advanced CNC Mill Programming and Applied Mathematics Level 2

Mill Project 304

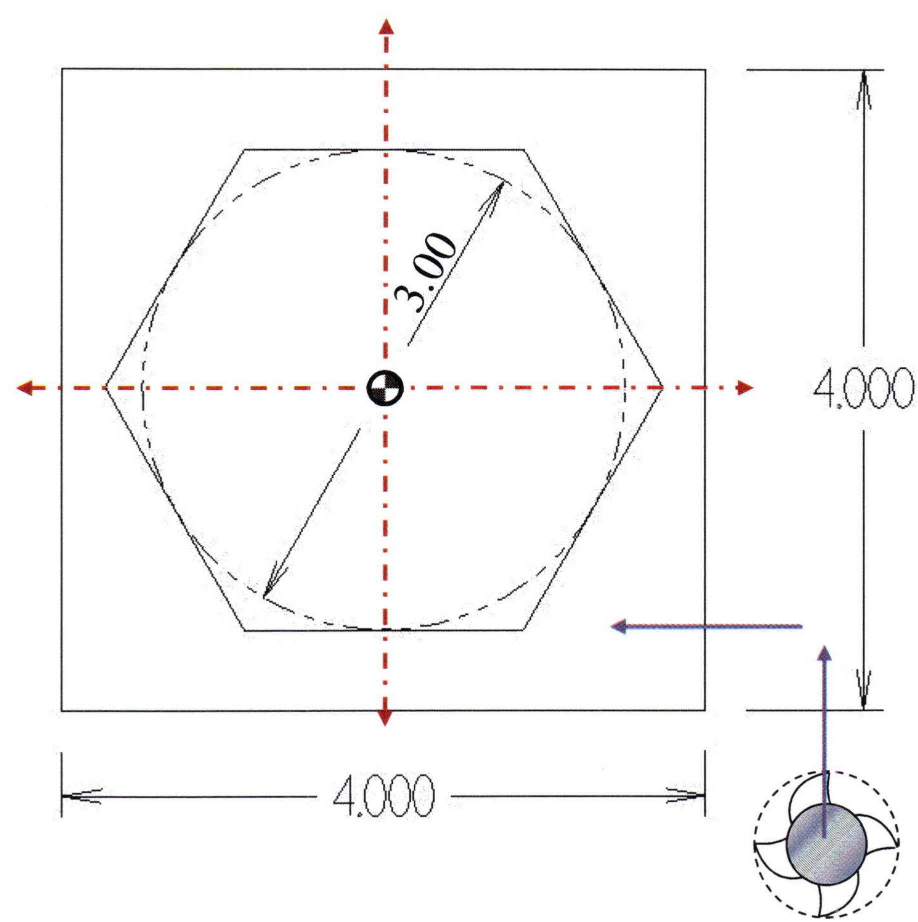

Advanced CNC Mill Programming and Applied Mathematics Level 2

A **regular polygon** is a polygon whose sides are all the same length and whose angles are all the same. In a regular hexagon, joining the ends of the sides to the center of the hexagon forms an equilateral triangle. An equilateral triangle is a triangle whose sides are equal in length. To calculate the CNC coordinates, the *side lengths* must be calculated.

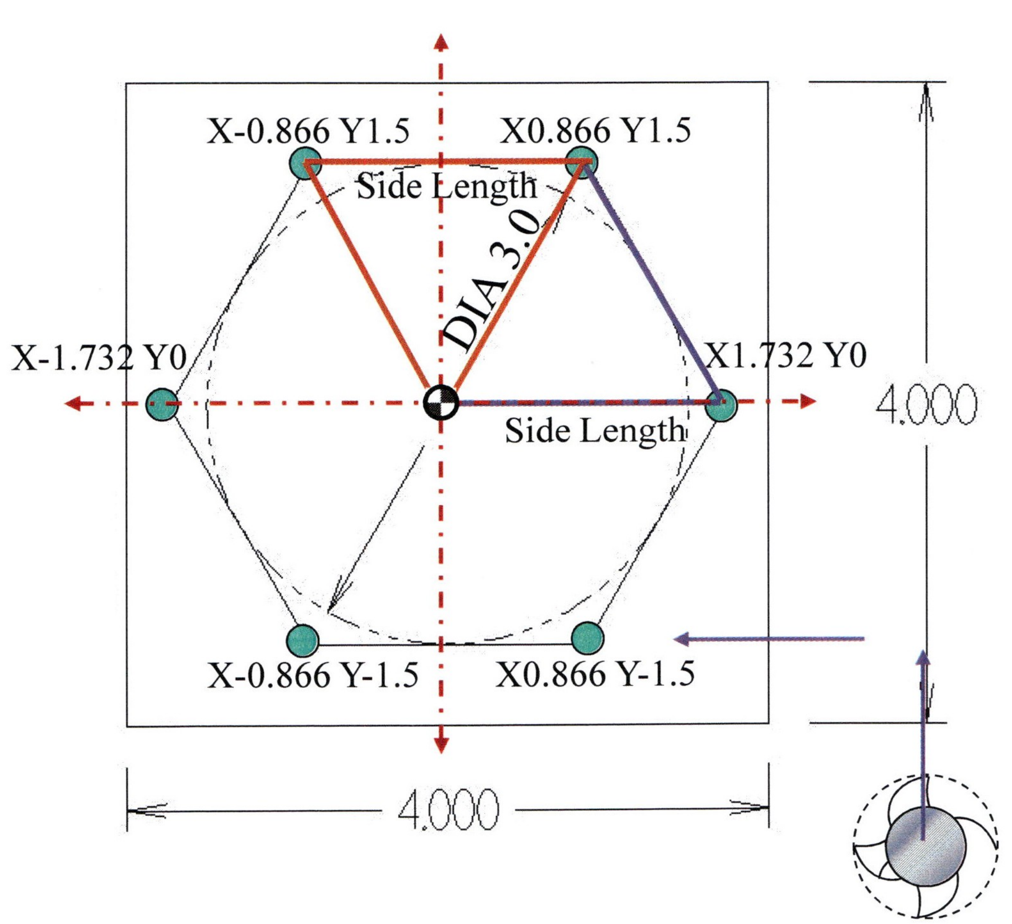

Step 1: Draw an equilateral triangle:

Step 2: Construct Right Triangle.
360 / 6 sides = 60 degrees

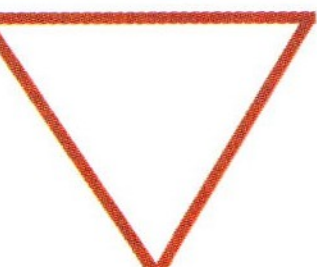

Step 3: Calculate X:

$$tan(60) = \frac{1.5}{adjacent}$$

$$1.73205 = \frac{1.5}{adjacent}$$

$$adjacent = \frac{1.5}{1.73205}$$

$$adjacent = 0.866$$

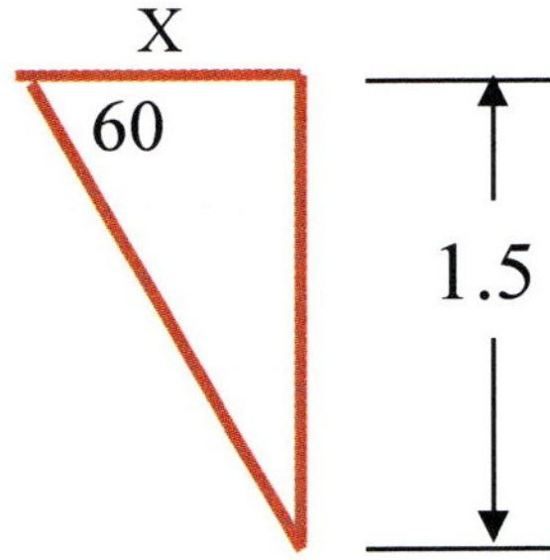

Mill Project 304 PROGRAM

```
%
O304 ( ORIGIN CENTER )
      ( TOOL ON )
G54 G90 G0 X2.0 Y-2.5 (START IN LR)
T2 M6
S1000 M3
G43 H2 Z0.1
G1 Z-0.1 F10.0 M8
G41 D2 Y-1.5
X-0.866
X-1.732 Y0
X-.866 Y1.5
X0.866
X1.732 Y0
X0.866 Y-1.5
Y-2.5
G40 X2.0
G28 G91 G0 Z0
M30
%
```


   Advanced CNC Mill Programming and Applied Mathematics Level 2

Mill Project 305

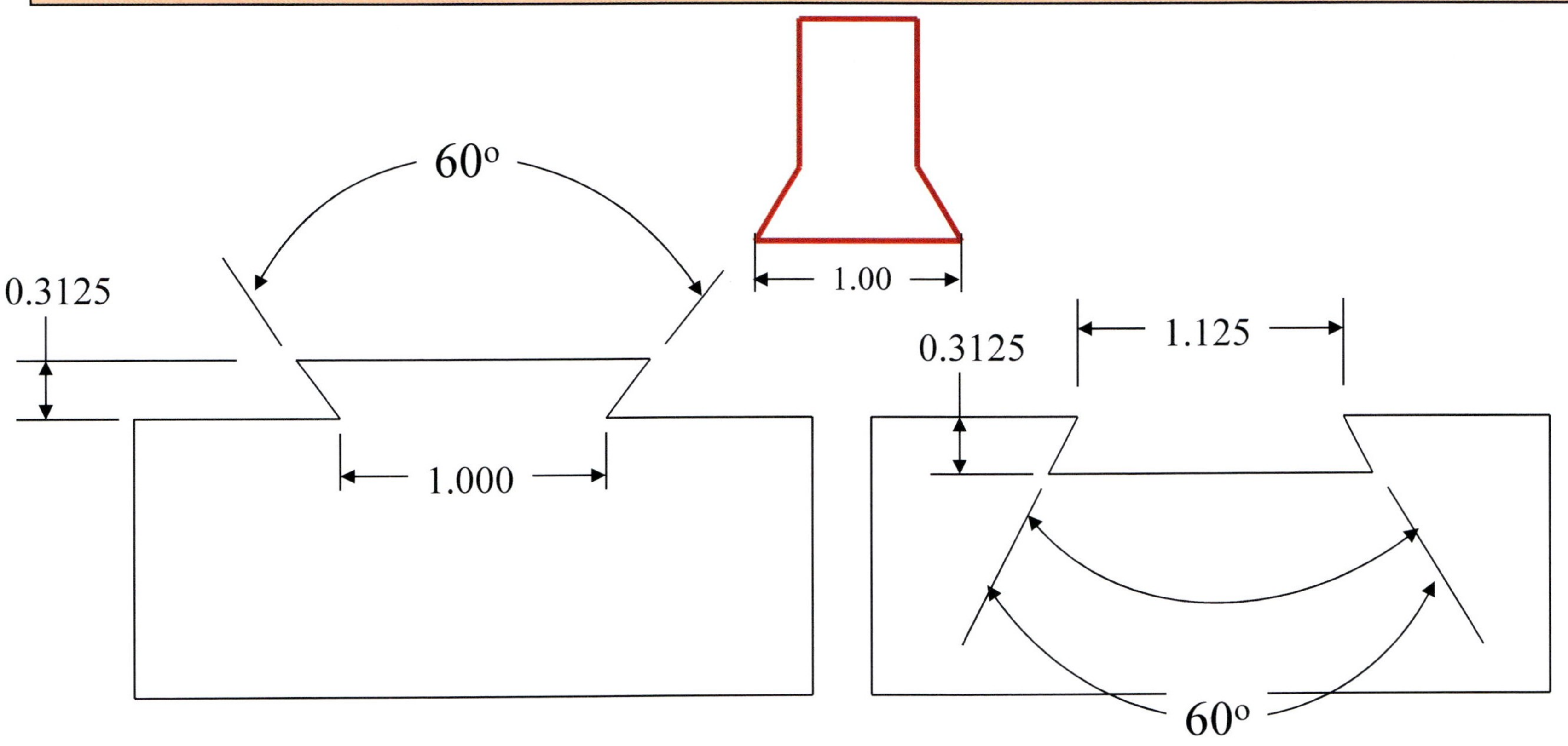

Advanced CNC Mill Programming and Applied Mathematics Level 2

305-1 Male Dovetail

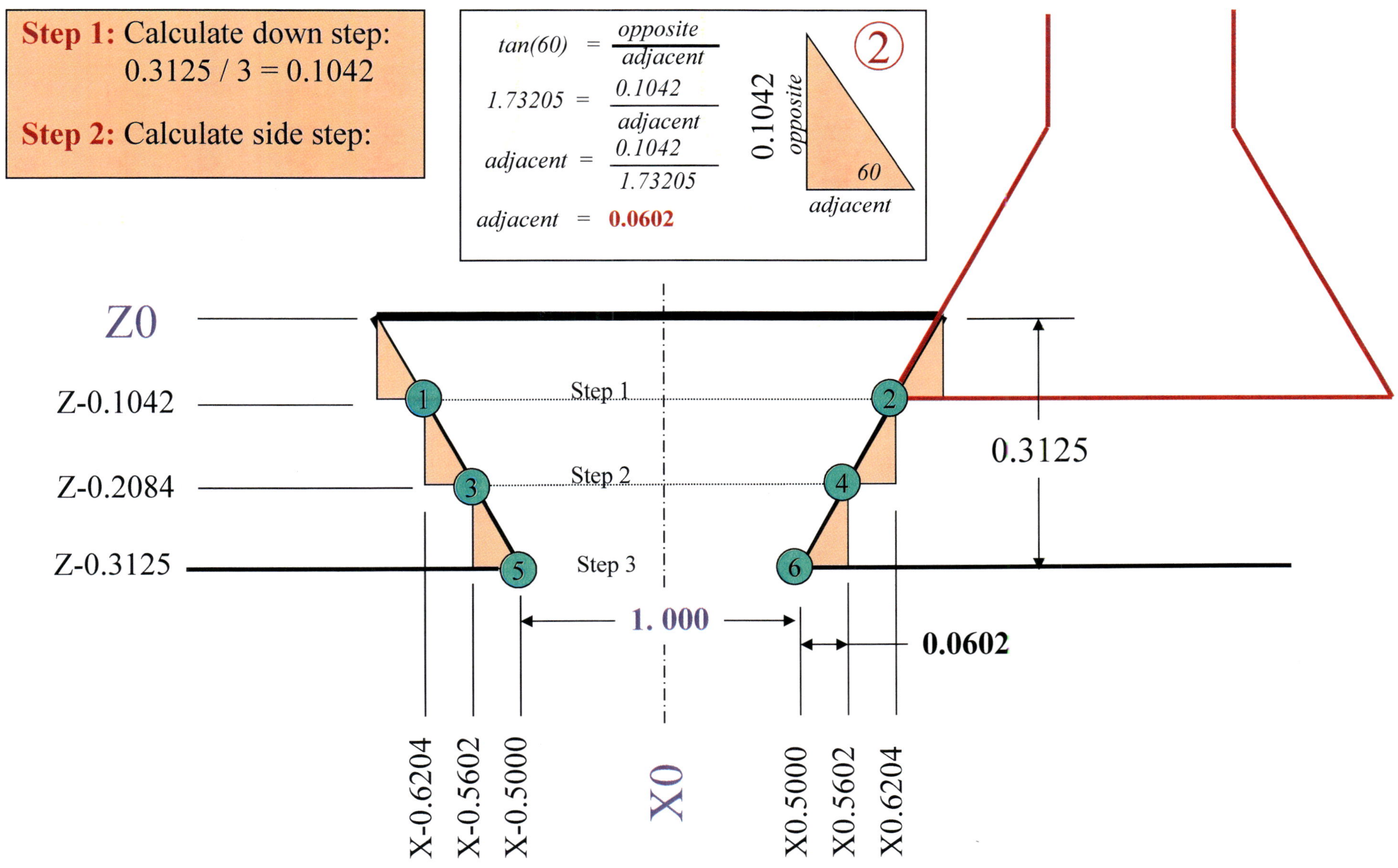

Advanced CNC Mill Programming and Applied Mathematics Level 2

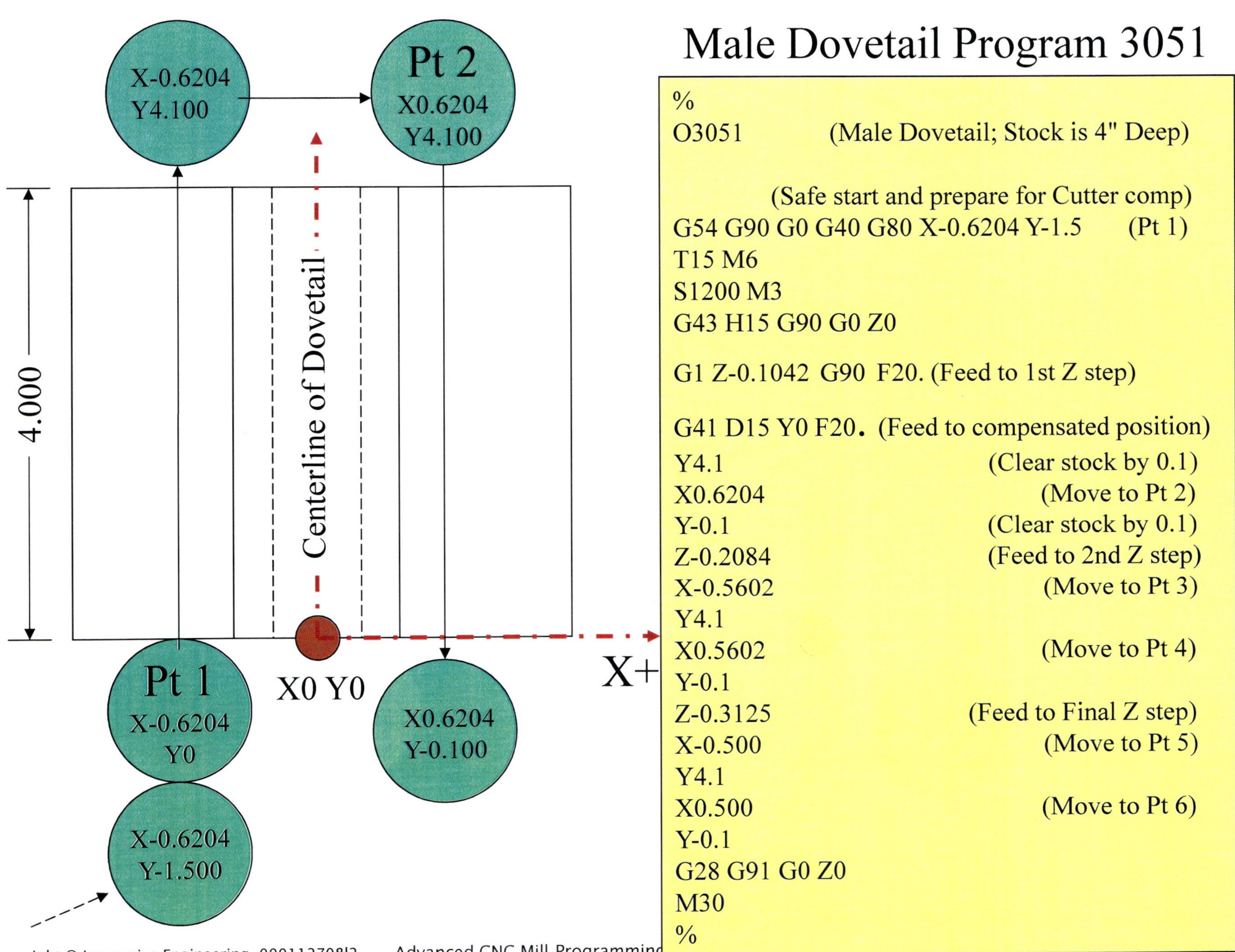

Copyright © Immersive Engineering 000112708I2   Advanced CNC Mill Programming

305-2 Female Dovetail

Using the same down and side steps, develop the female program.

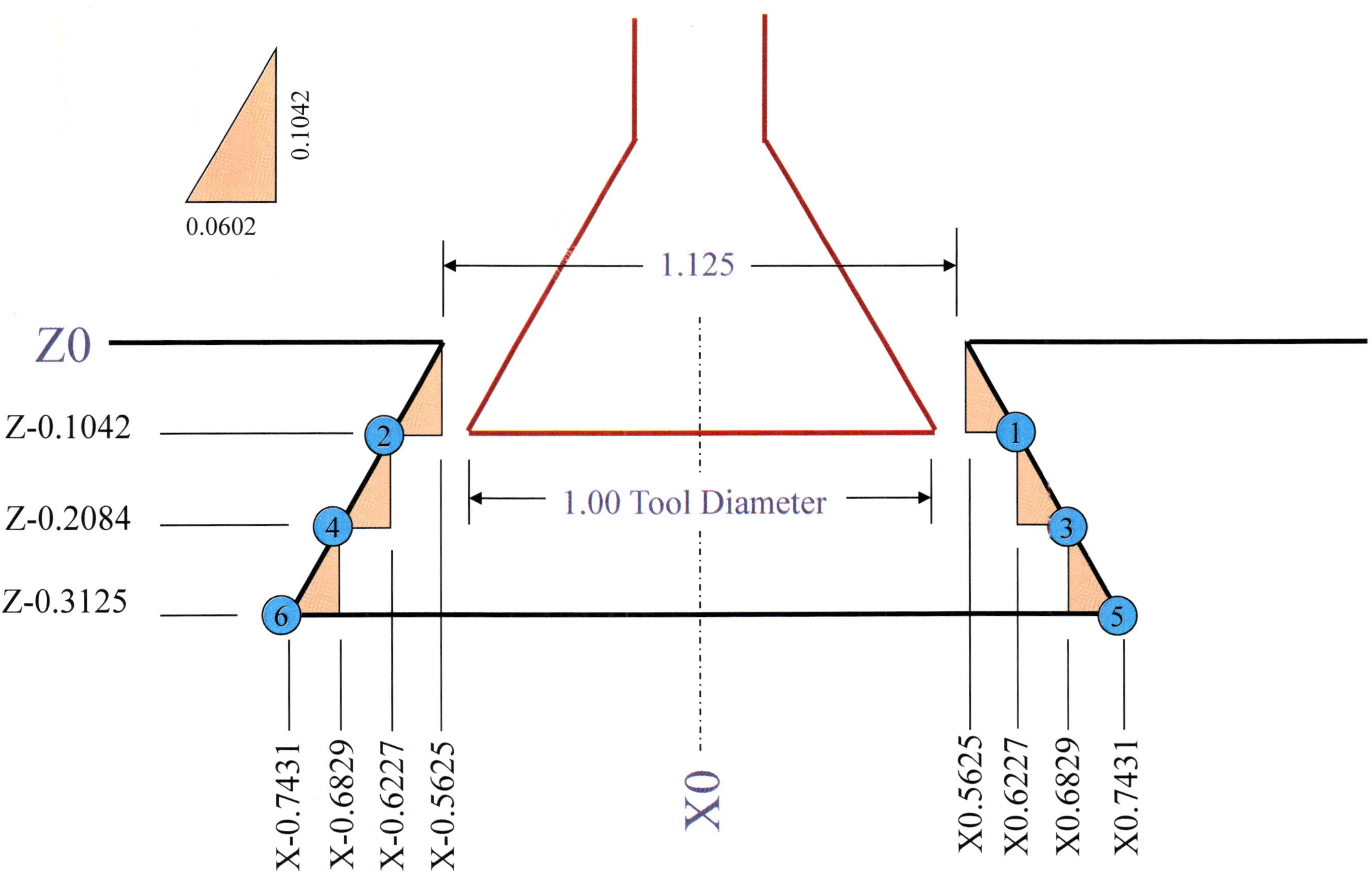

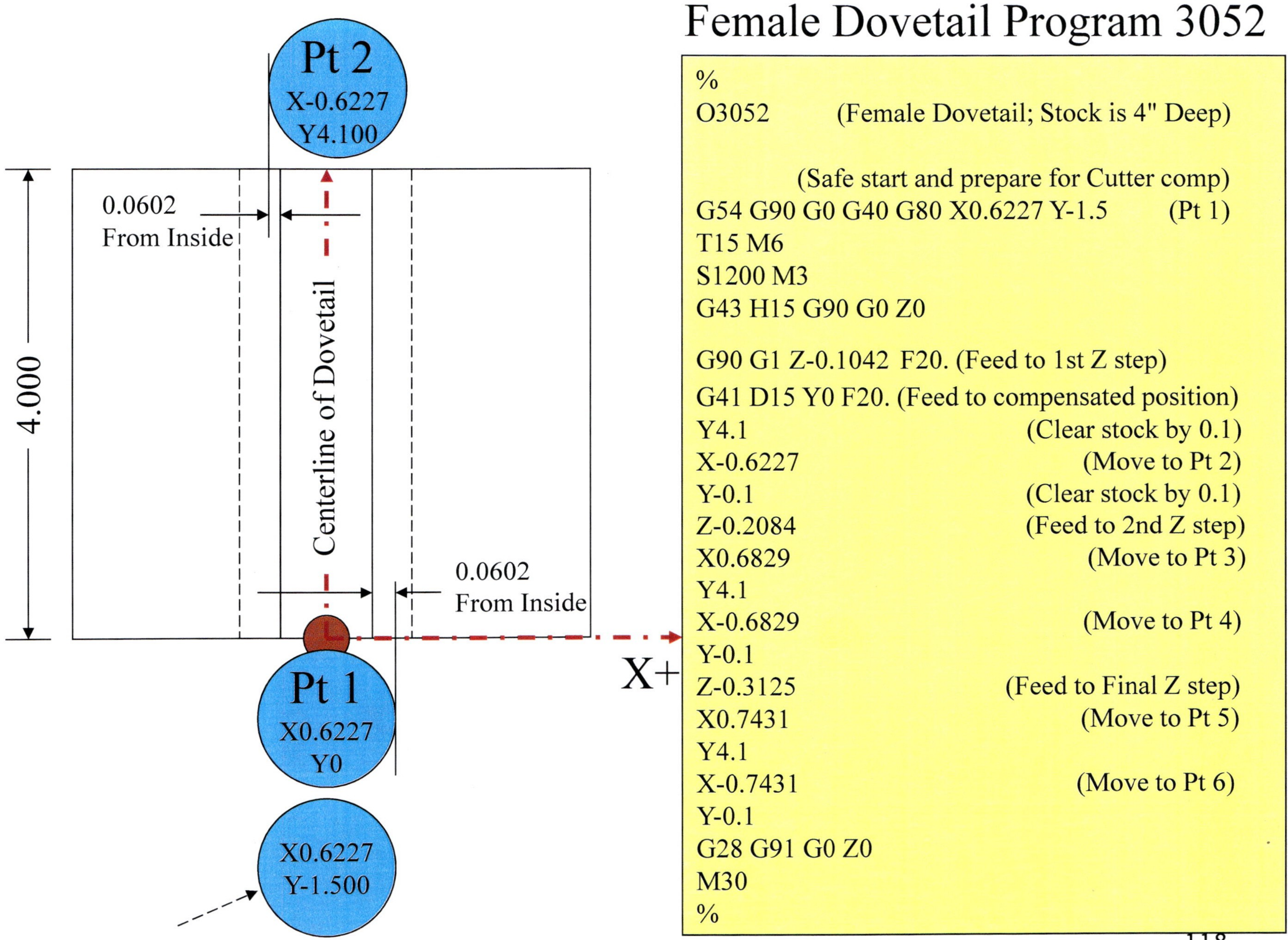

Female Dovetail Program 3052

Pt 2
X-0.6227
Y4.100

Pt 1
X0.6227
Y0

X0.6227
Y-1.500

0.0602
From Inside

0.0602
From Inside

4.000

Centerline of Dovetail

X+

%
O3052   (Female Dovetail; Stock is 4" Deep)

               (Safe start and prepare for Cutter comp)
G54 G90 G0 G40 G80 X0.6227 Y-1.5   (Pt 1)
T15 M6
S1200 M3
G43 H15 G90 G0 Z0

G90 G1 Z-0.1042 F20. (Feed to 1st Z step)
G41 D15 Y0 F20. (Feed to compensated position)
Y4.1   (Clear stock by 0.1)
X-0.6227   (Move to Pt 2)
Y-0.1   (Clear stock by 0.1)
Z-0.2084   (Feed to 2nd Z step)
X0.6829   (Move to Pt 3)
Y4.1
X-0.6829   (Move to Pt 4)
Y-0.1
Z-0.3125   (Feed to Final Z step)
X0.7431   (Move to Pt 5)
Y4.1
X-0.7431   (Move to Pt 6)
Y-0.1
G28 G91 G0 Z0
M30
%

Mill Project 401
Machine the inside of the track using 3 profiles.

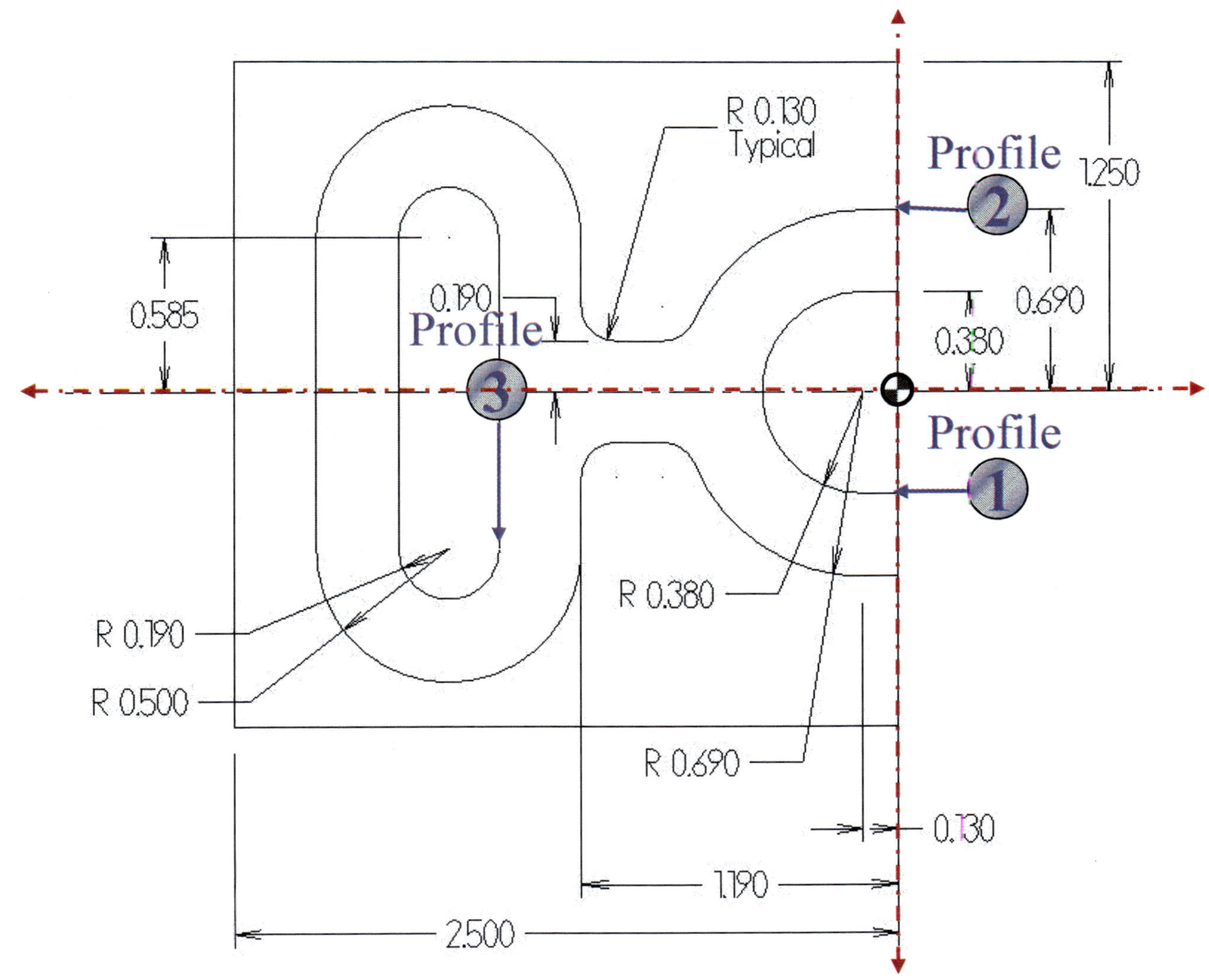

Advanced CNC Mill Programming and Applied Mathematics Level 2

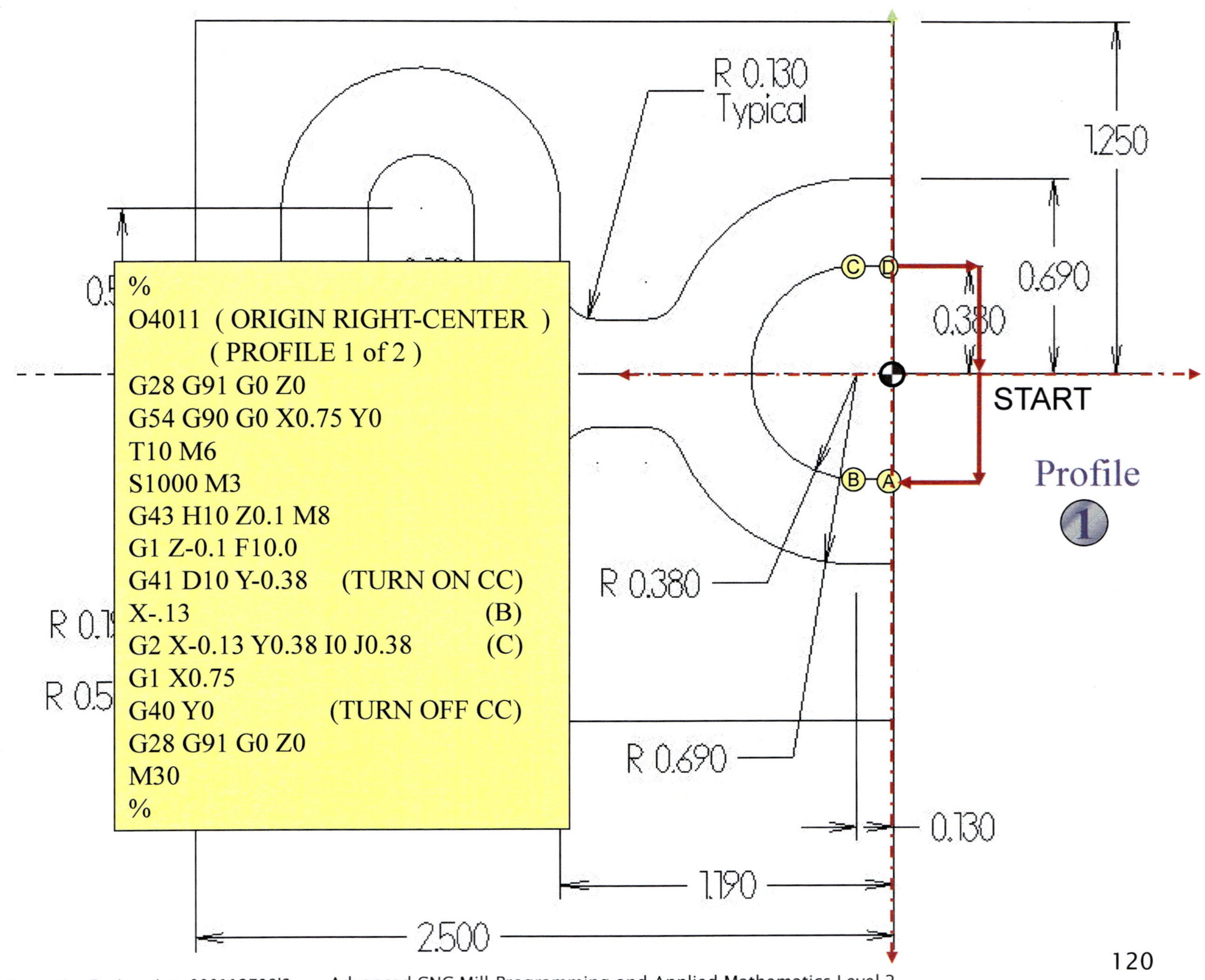
R 0.130
Typical
1.250
C D
0.690
0.380
START
Profile
1
%
O4011 (ORIGIN RIGHT-CENTER)
      (PROFILE 1 of 2)
G28 G91 G0 Z0
G54 G90 G0 X0.75 Y0
T10 M6
S1000 M3
G43 H10 Z0.1 M8
G1 Z-0.1 F10.0
G41 D10 Y-0.38   (TURN ON CC)
X-.13   (B)
G2 X-0.13 Y0.38 I0 J0.38   (C)
G1 X0.75
G40 Y0   (TURN OFF CC)
G28 G91 G0 Z0
M30
%
B A
R 0.380
R 0.690
0.130
1.190
2.500

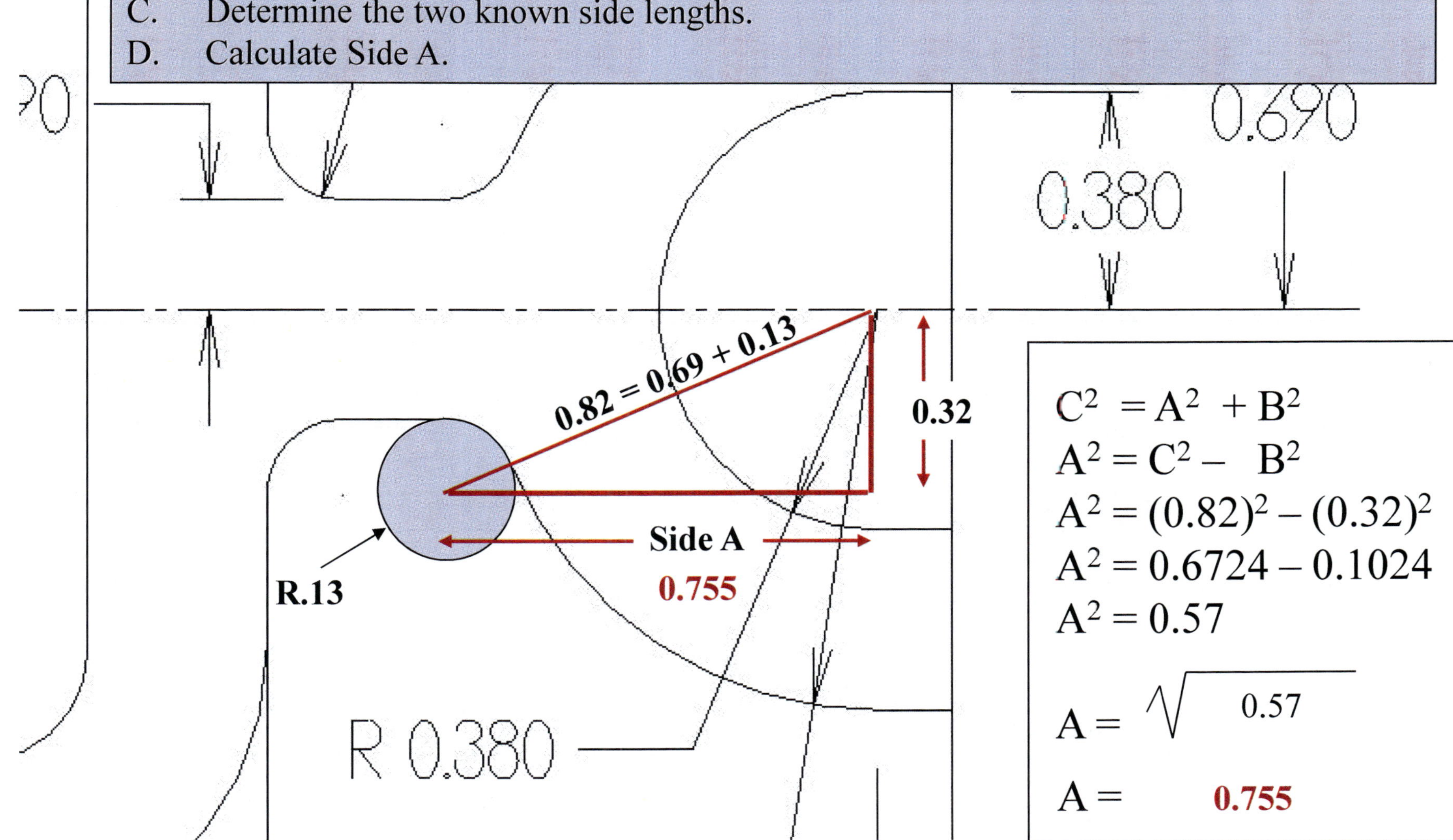

Profile 2 requires the use of geometry and right angle trig as shown.
A.   Draw an angled line from center of 0.13 radius to center of 0.38 radius.
B.   Draw a horizontal line from the center of the 0.13 radius to the 0.38 radius.
C.   Determine the two known side lengths.
D.   Calculate Side A.
0.690
0.380
0.82 = 0.69 + 0.13
0.32
Side A
0.755
R.13
R 0.380
$C^2 = A^2 + B^2$
$A^2 = C^2 - B^2$
$A^2 = (0.82)^2 - (0.32)^2$
$A^2 = 0.6724 - 0.1024$
$A^2 = 0.57$
$A = \sqrt{0.57}$
$A = 0.755$

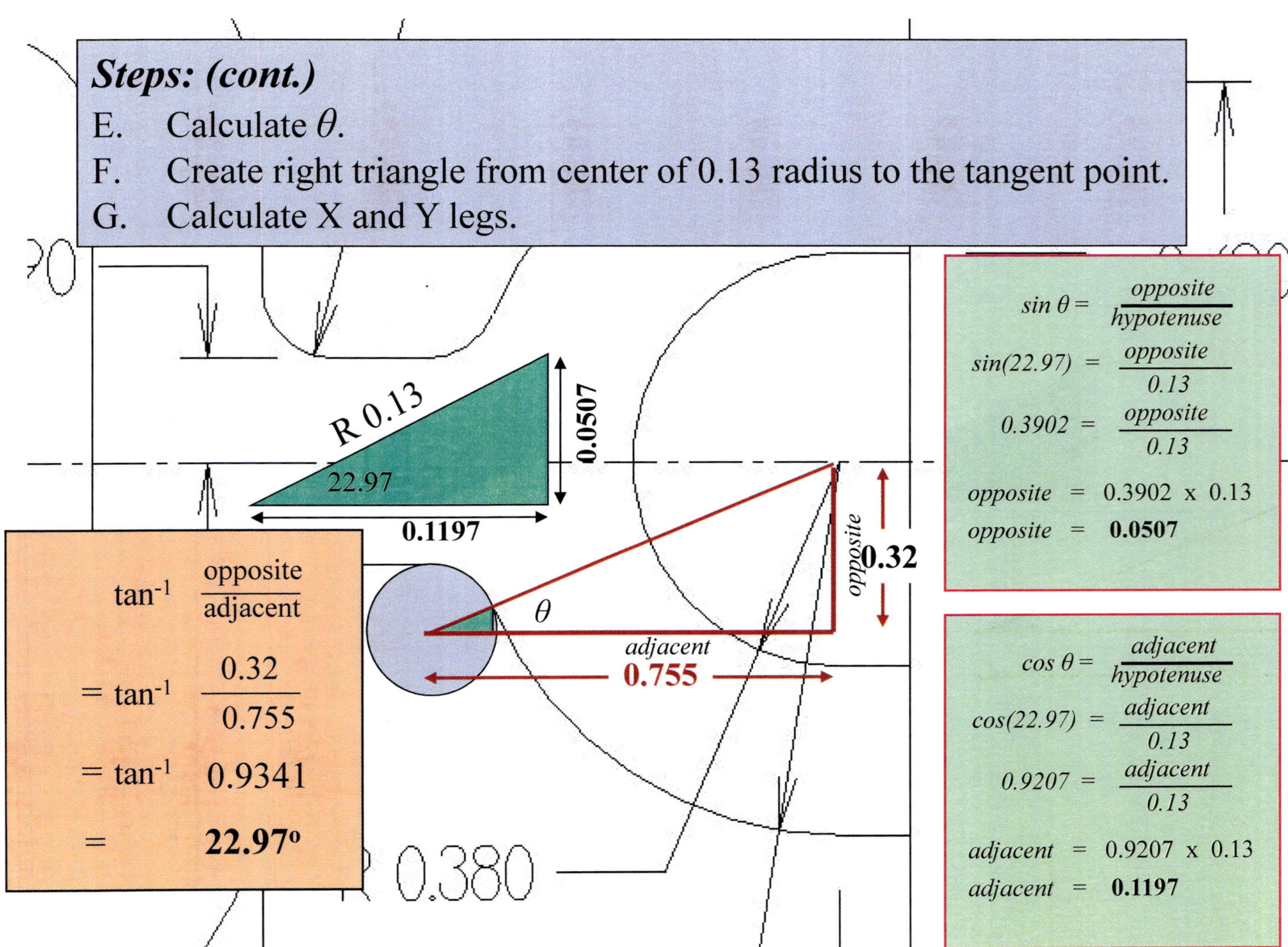

Steps: (cont.)
E.   Calculate θ.
F.   Create right triangle from center of 0.13 radius to the tangent point.
G.   Calculate X and Y legs.

R 0.13
22.97
0.0507
0.1197

θ
opposite
0.32
adjacent
0.755

$\tan^{-1} \dfrac{opposite}{adjacent}$

$= \tan^{-1} \dfrac{0.32}{0.755}$

$= \tan^{-1} \ 0.9341$

$= \ \mathbf{22.97^o}$

$\sin \theta = \dfrac{opposite}{hypotenuse}$

$\sin(22.97) = \dfrac{opposite}{0.13}$

$0.3902 = \dfrac{opposite}{0.13}$

$opposite = 0.3902 \times 0.13$

$opposite = \mathbf{0.0507}$

$\cos \theta = \dfrac{adjacent}{hypotenuse}$

$\cos(22.97) = \dfrac{adjacent}{0.13}$

$0.9207 = \dfrac{adjacent}{0.13}$

$adjacent = 0.9207 \times 0.13$

$adjacent = \mathbf{0.1197}$

0.380

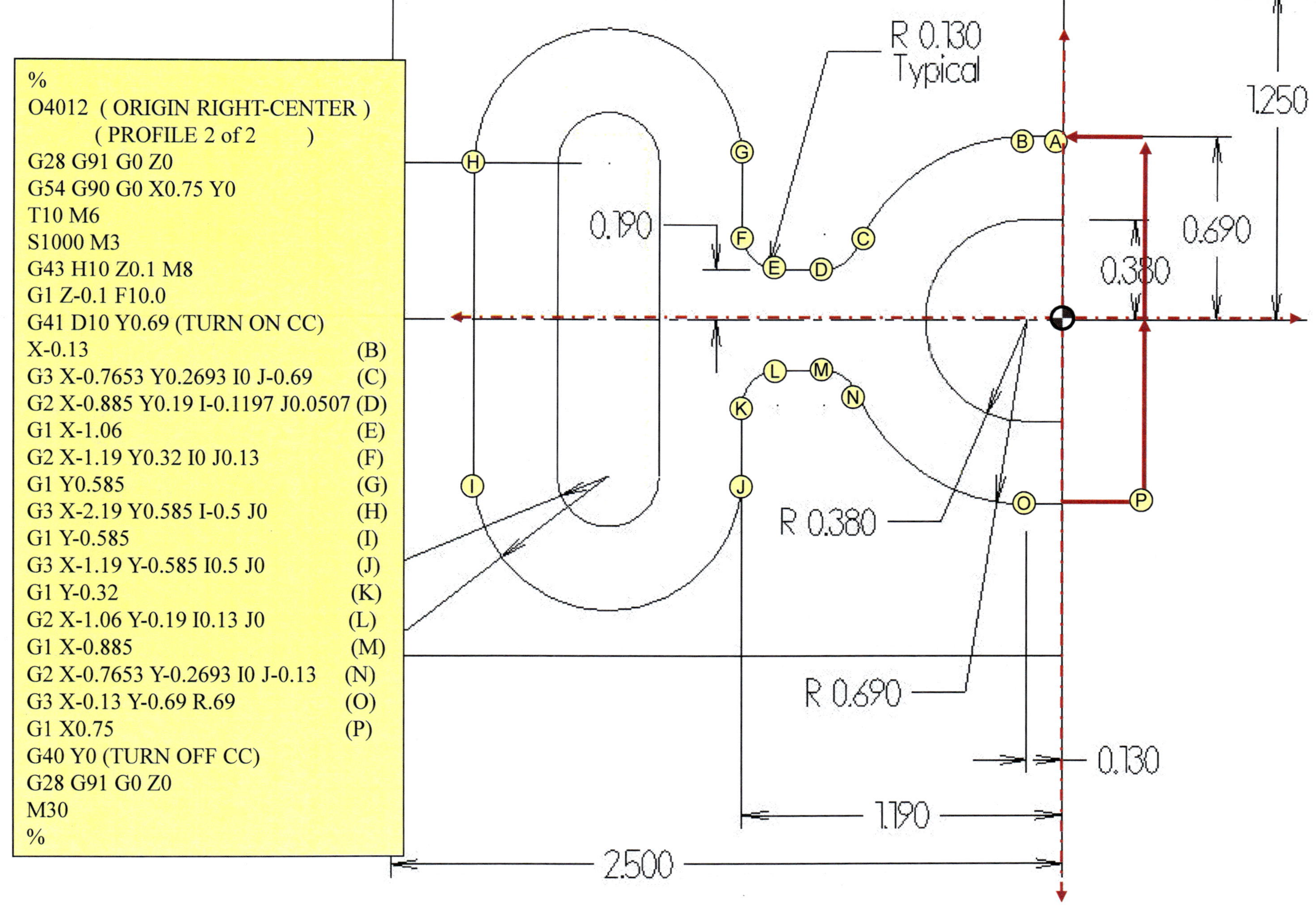

%
O4012 (ORIGIN RIGHT-CENTER)
       (PROFILE 2 of 2   )
G28 G91 G0 Z0
G54 G90 G0 X0.75 Y0
T10 M6
S1000 M3
G43 H10 Z0.1 M8
G1 Z-0.1 F10.0
G41 D10 Y0.69 (TURN ON CC)
X-0.13   (B)
G3 X-0.7653 Y0.2693 I0 J-0.69   (C)
G2 X-0.885 Y0.19 I-0.1197 J0.0507 (D)
G1 X-1.06   (E)
G2 X-1.19 Y0.32 I0 J0.13   (F)
G1 Y0.585   (G)
G3 X-2.19 Y0.585 I-0.5 J0   (H)
G1 Y-0.585   (I)
G3 X-1.19 Y-0.585 I0.5 J0   (J)
G1 Y-0.32   (K)
G2 X-1.06 Y-0.19 I0.13 J0   (L)
G1 X-0.885   (M)
G2 X-0.7653 Y-0.2693 I0 J-0.13   (N)
G3 X-0.13 Y-0.69 R.69   (O)
G1 X0.75   (P)
G40 Y0 (TURN OFF CC)
G28 G91 G0 Z0
M30
%
R 0.130
Typical
1.250
0.690
0.380
0.190
R 0.380
R 0.690
0.130
1.190
2.500
H
G
B
A
F
E
D
C
L
M
N
K
I
J
O
P

Mill Project 402

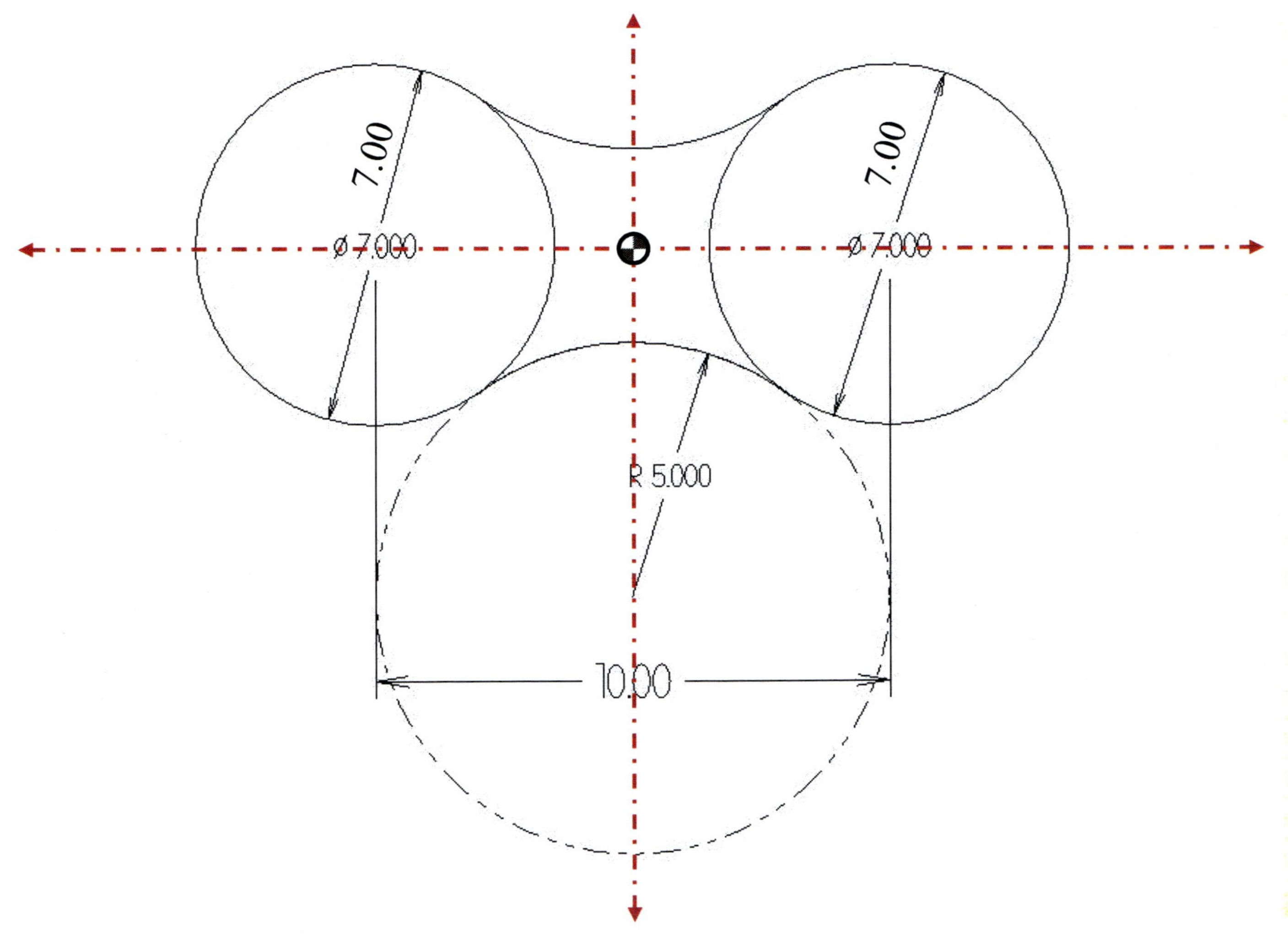

Advanced CNC Mill Programming and Applied Mathematics Level 2

Mill Project 402

Step 1: Construct main right triangle and solve for Side A.

$$C^2 = A^2 + B^2$$
$$A^2 = C^2 - B^2$$
$$A^2 = (8.5)^2 - (5)^2$$
$$A^2 = 72.5 - 25$$
$$A^2 = 47.25$$

$$A = \sqrt{47.25}$$

$$A = \mathbf{6.8739}$$

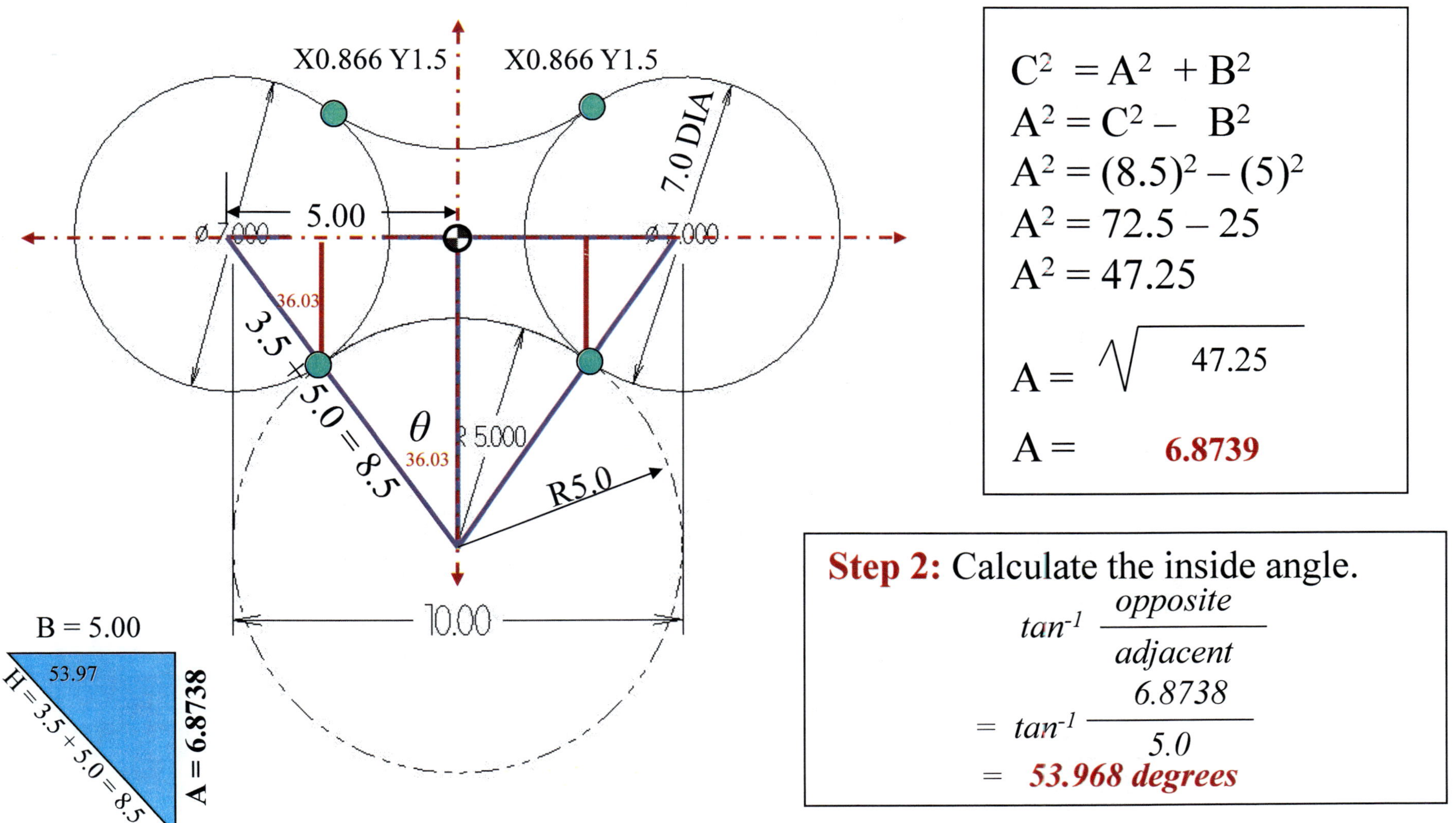

Step 2: Calculate the inside angle.

$$tan^{-1} \frac{opposite}{adjacent}$$
$$= tan^{-1} \frac{6.8738}{5.0}$$
$$= \mathbf{53.968\ degrees}$$

     Advanced CNC Mill Programming and Applied Mathematics Level 2

Mill Project 402

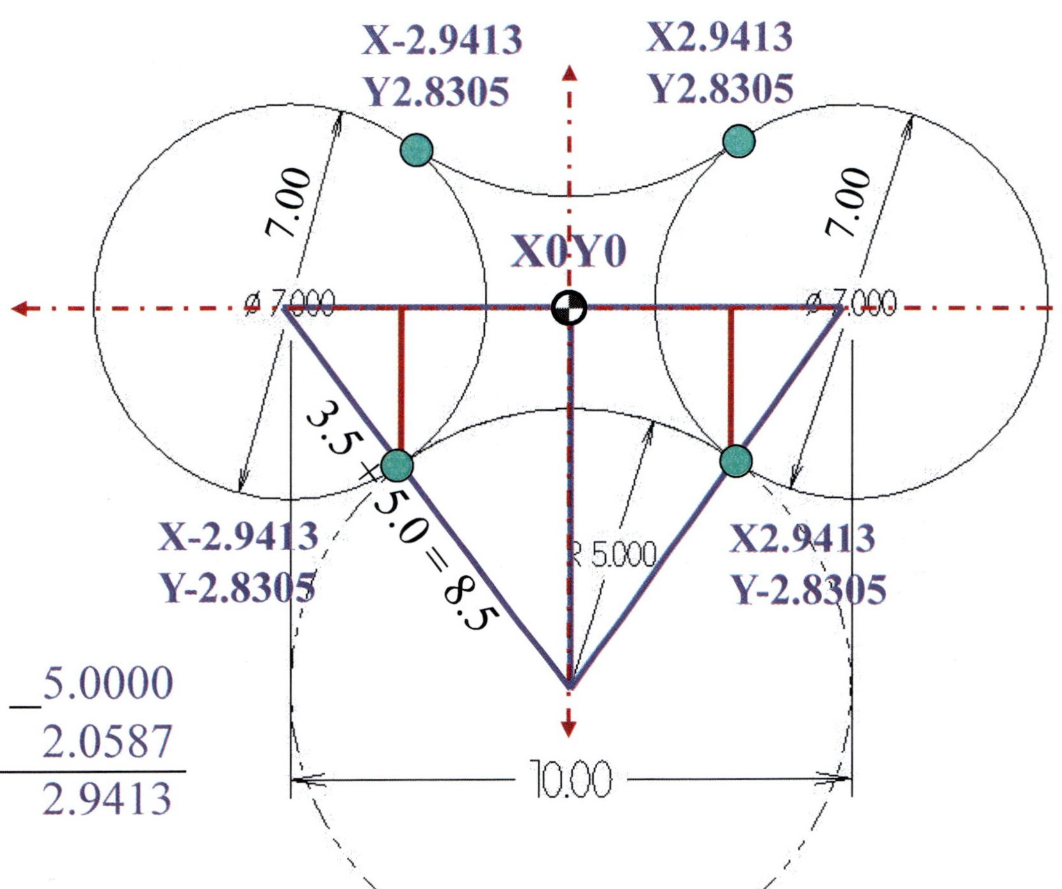

Step 3a: Calculate leg lengths using θ and the hypotenuse.

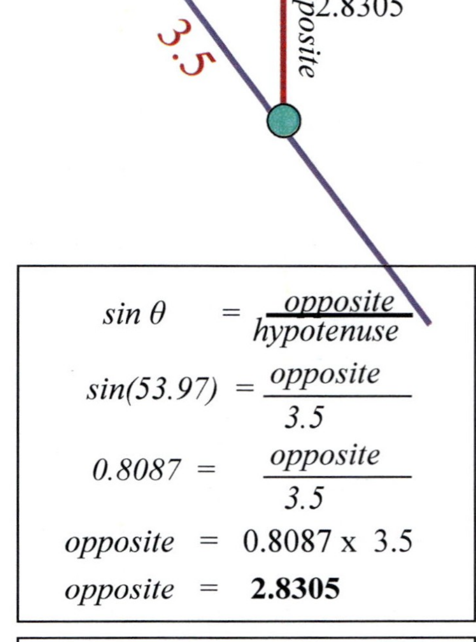

$$sin\ \theta = \frac{opposite}{hypotenuse}$$

$$sin(53.97) = \frac{opposite}{3.5}$$

$$0.8087 = \frac{opposite}{3.5}$$

$$opposite = 0.8087 \times 3.5$$

$$opposite = \mathbf{2.8305}$$

$$cos\ \theta = \frac{adjacent}{hypotenuse}$$

$$cos(53.97) = \frac{adjacent}{3.5}$$

$$0.5882 = \frac{adjacent}{3.5}$$

$$adjacent = 0.5882 \times 3.5$$

$$adjacent = \mathbf{2.0587}$$

Step 3b: Use the leg lengths to calculate the XY coordinates at each tangent point.

Mill Project 402

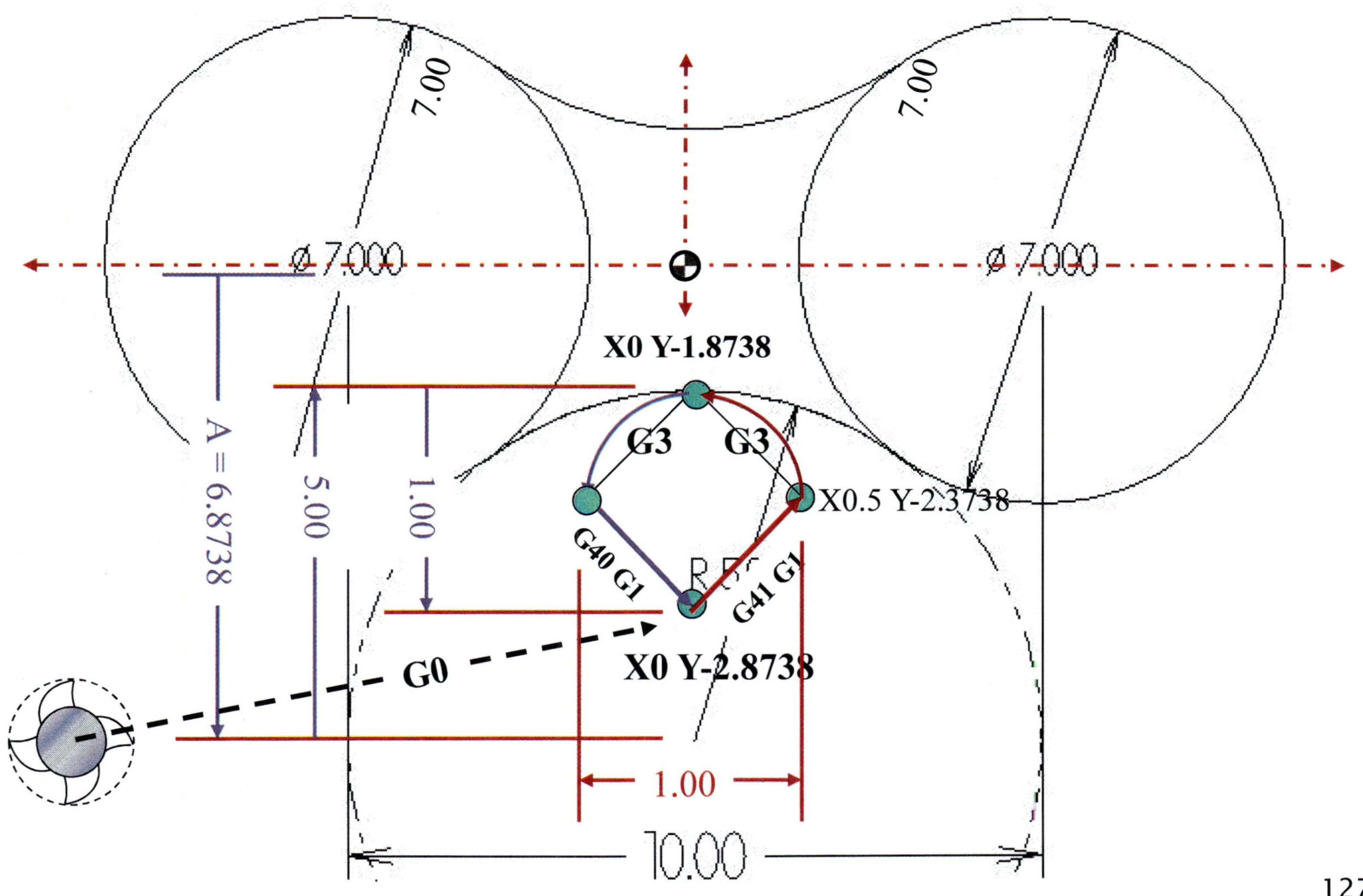

Advanced CNC Mill Programming and Applied Mathematics Level 2

Mill Project 402 PROGRAM using combination of R and IJ for radii

```
%
O402 ( Outer profile )
G54 G90 G0 X0 Y-2.8738 (Move to Home base position)
T2 M6
S1000 M3
G43 Z0.1
G1 H2 Z-0.1 F10. M8
G41 D2 X0.5 Y-2.3738        (1ST Base)
G3 X0 Y-1.8738 R0.5         (Arc into 2ND base)
G3 X-2.9413 Y-2.8305 R5.0
G2 X-2.9413 Y2.8305 I-2.0587 J2.8305
G3 X2.9413 Y2.8305 R5.0
G2 X2.9413 Y-2.8305 I2.0587 J-2.8305
G3 X0 Y-1.8738 R5.0         (2nd Base)
G3 X-0.5 Y-2.3738 R0.5      (Arc off to 3rd Base)
G40 G1 X0 Y-2.8738
G28 G91 G0 Z0
M30
%
```

Mill Project 403

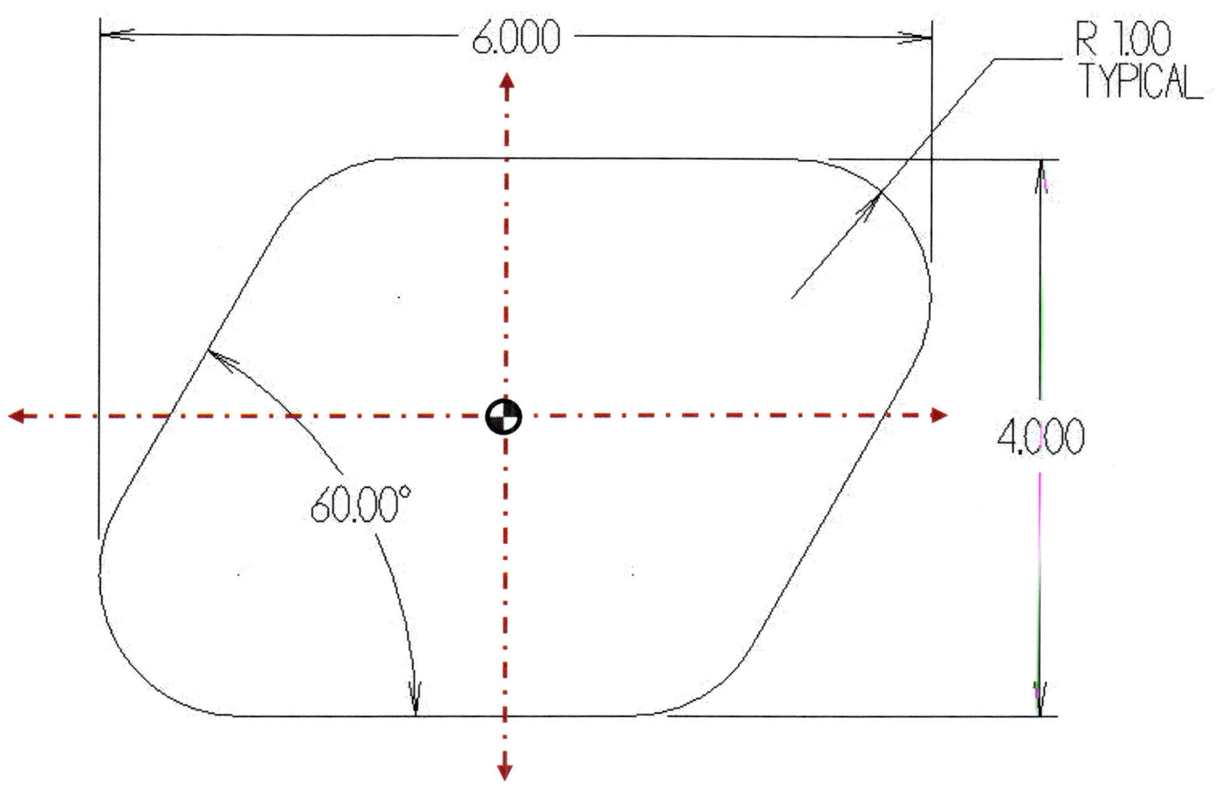

Advanced CNC Mill Programming and Applied Mathematics Level 2

Mill Project 403

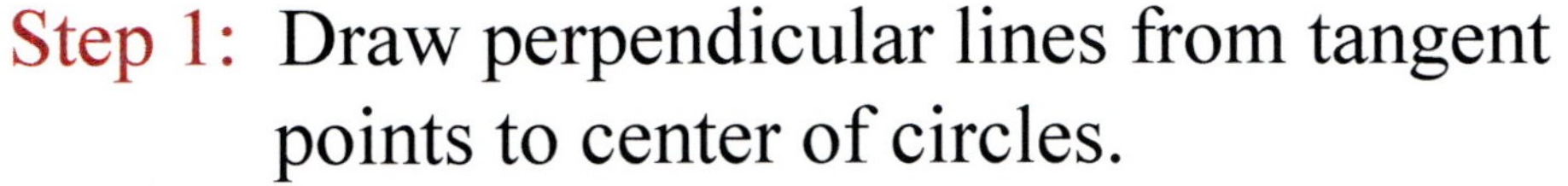

Trig Functions

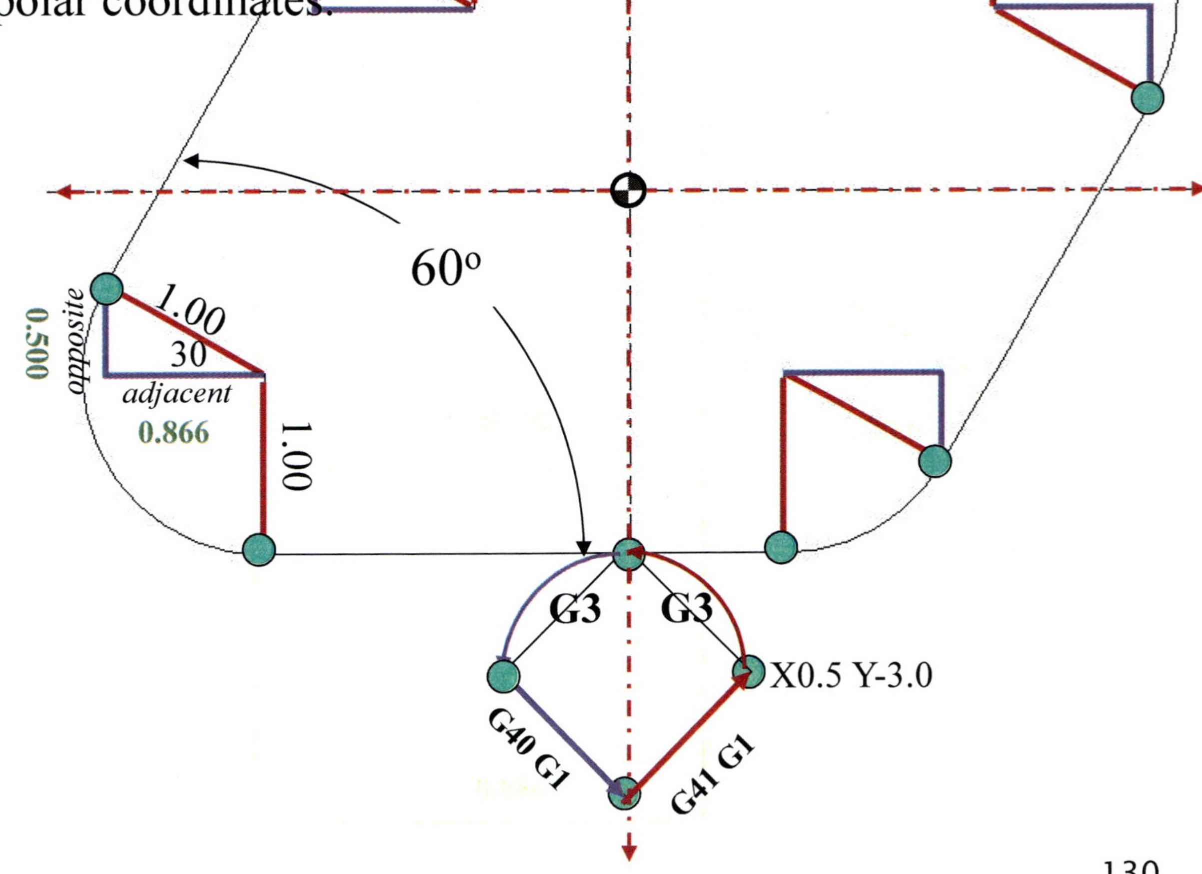

Advanced CNC Mill Programming and Applied Mathematics Level 2

Mill Project 403

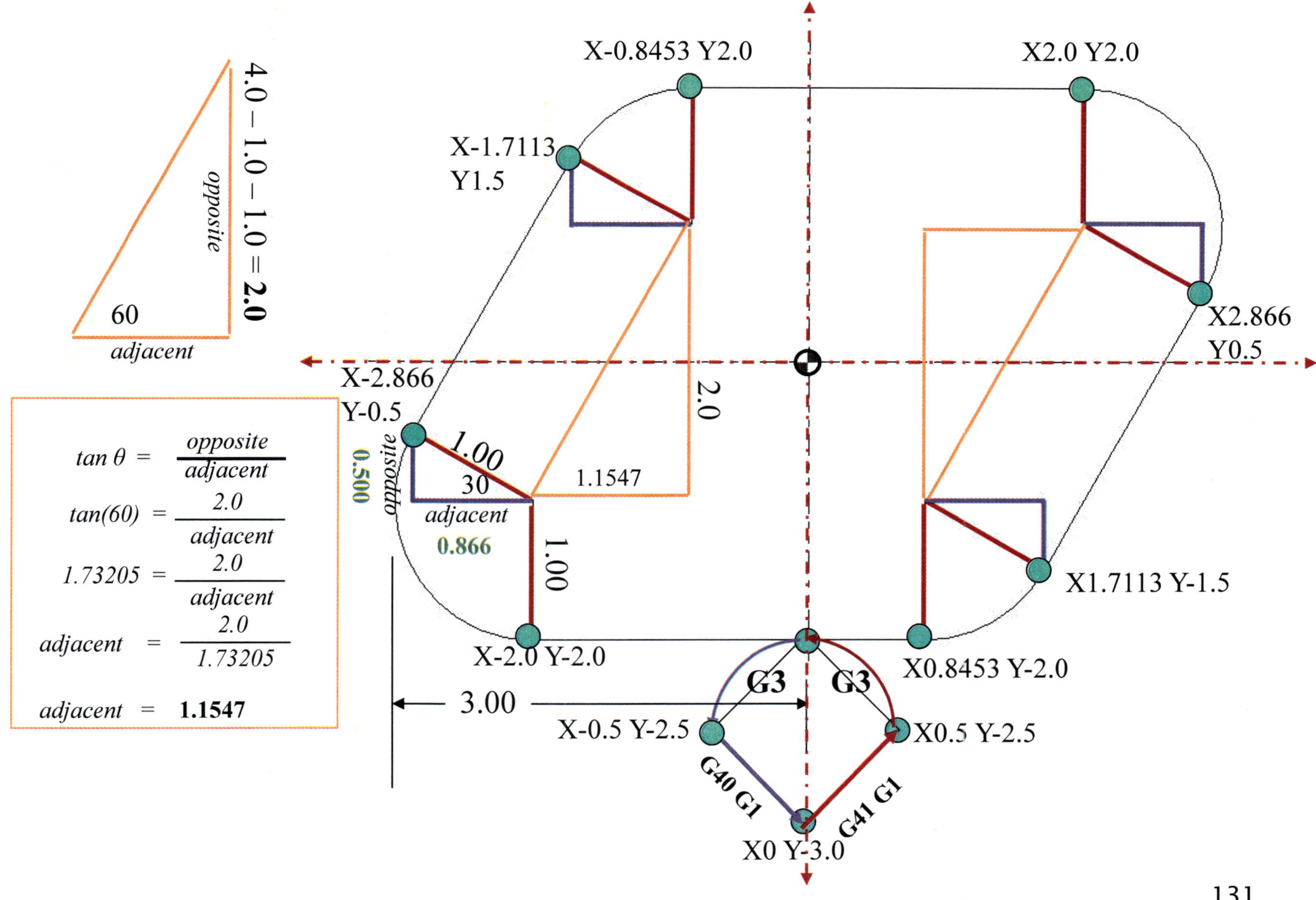

Mill Project 403 PROGRAM

```
%
O403 ( ORIGIN CENTER  TOOL ON )
G54 G90 G0 X0 Y-3.0
T2 M6
S1000 M3
G43 H2 Z0.1
G1 Z-.1 F10.0 M8
G41 D2 X0.5 Y-2.5 (1ST BASE)
G3 X0 Y-2.0 R0.5 (2ND BASE)
G1 X-2.0
G2 X-2.866 Y-0.5 R1.0 (LL)
G1 X-1.7113 Y1.5
G2 X-0.8453 Y2.0 R1.0 (UR)
G1 X2.0
G2 X2.866 Y0.5 R1.0 (UR)
G1 X1.7113 Y-1.5
G2 X0.8453 Y-2.0 R1.0 (LR)
G1 X0
G3 X-0.5 Y-2.5 R0.5 (ARC OFF)
G40 G1 X0 Y-3.0
G28 G91 G0 Z0
M30
%
```

Mill Project 404

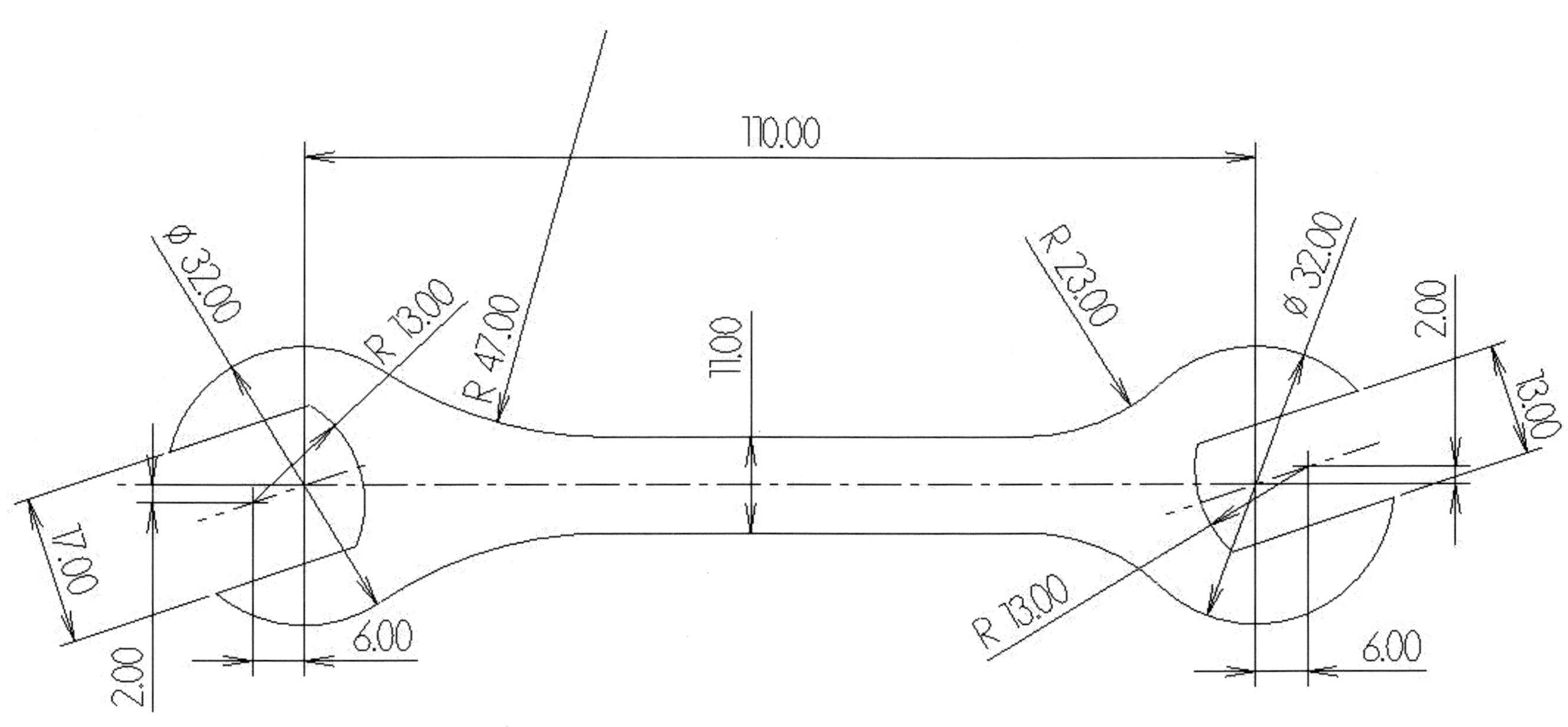

Mill Project 404-1 (Outside Only)

Steps:

A. Draw a line from tangent point of the line to the center of the 47mm radius.

B. Draw a line from the center of the 16mm radius to the center of the 47mm radius

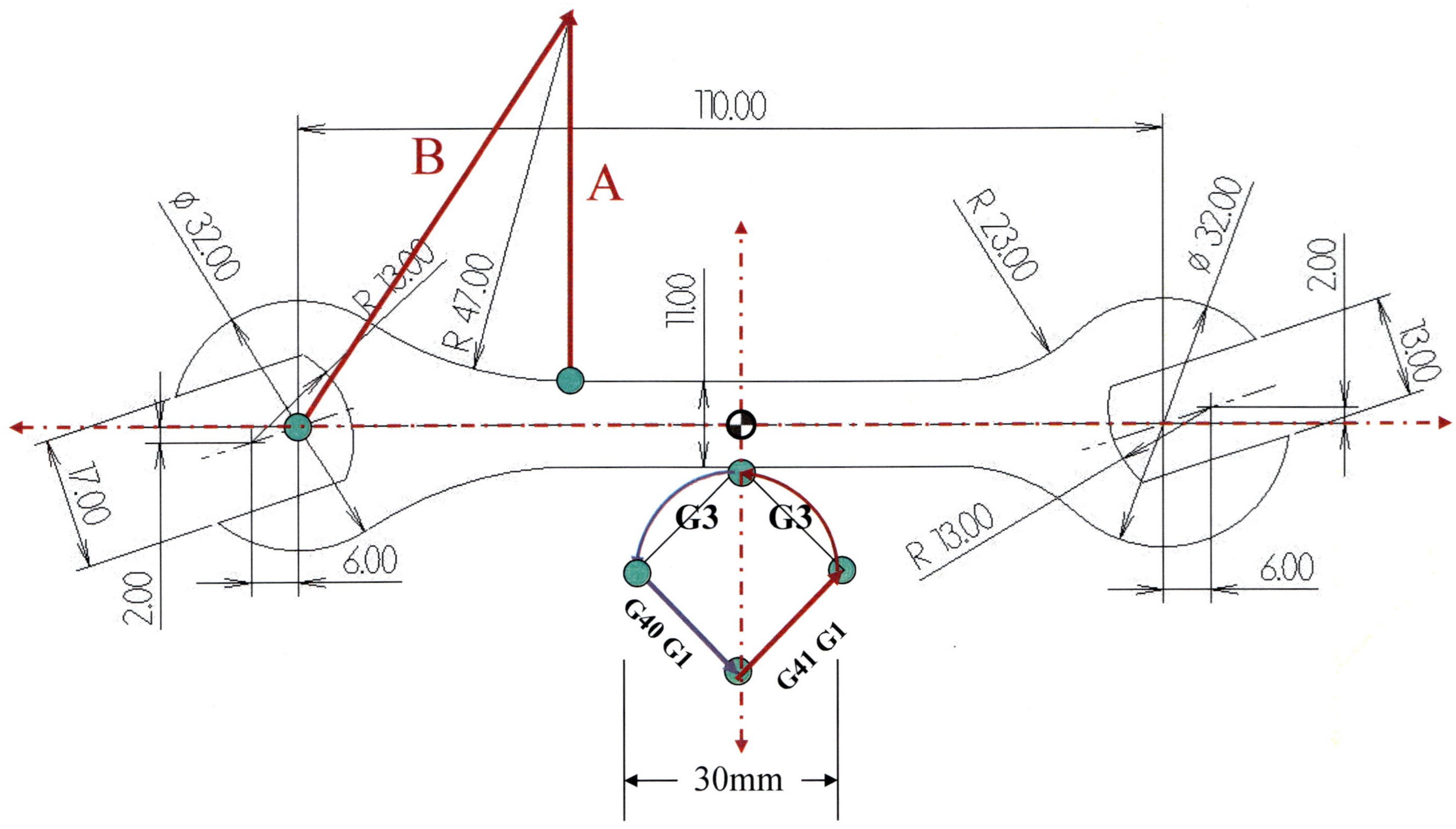

    Advanced CNC Mill Programming and Applied Mathematics Level 2

C. Draw a horizontal line to form a right triangle.

D. Extend the vertical line down to the intersection.

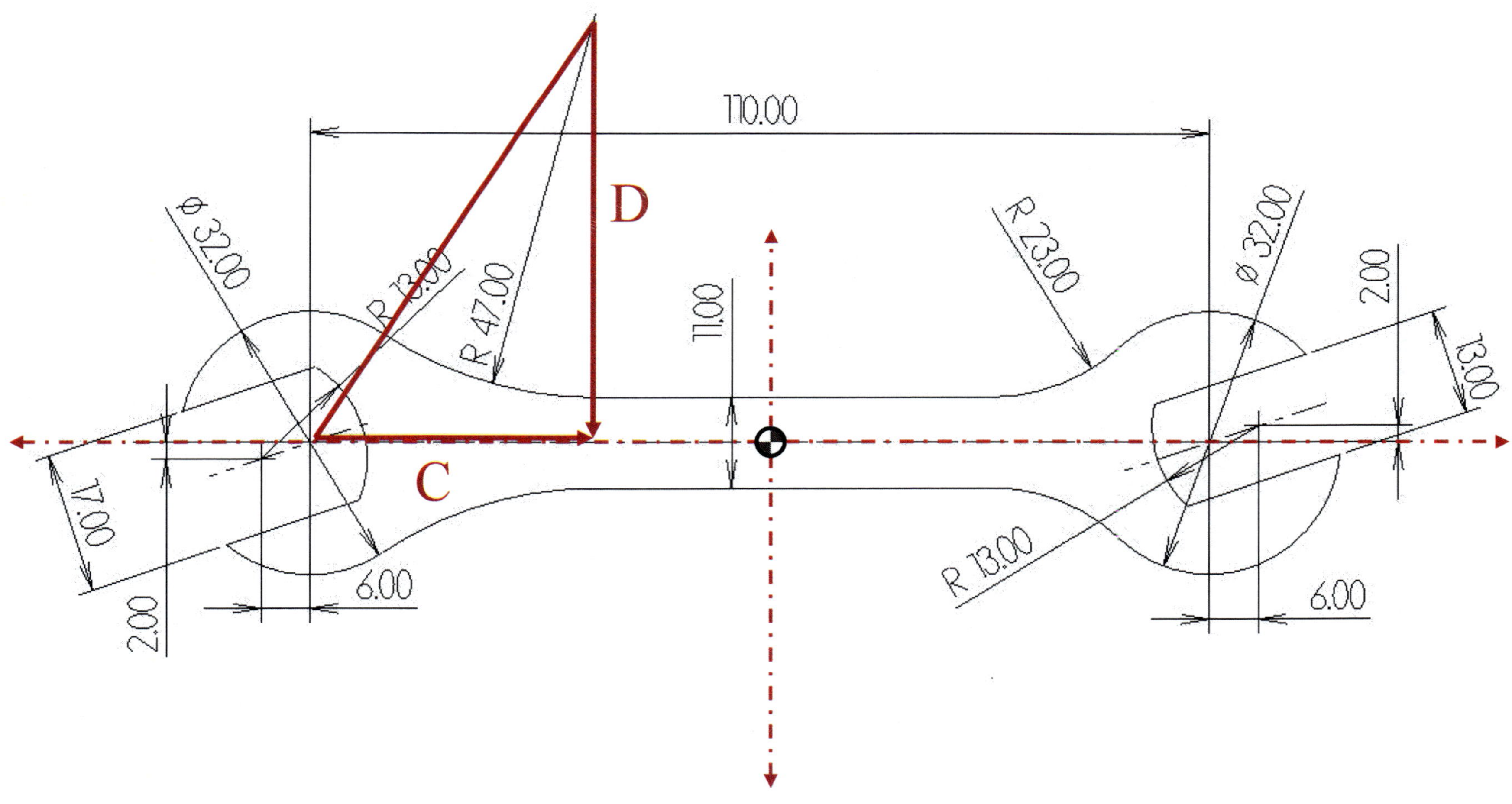

E.   Calculate side A.

F.   Calculate θ.

$$A^2 = C^2 - B^2$$

$$A^2 = 63^2 - 52.5^2$$

$$A^2 = 3969 - 2756.25$$

$$A = \sqrt{1212.75}$$

$$A = \mathbf{34.8245}$$

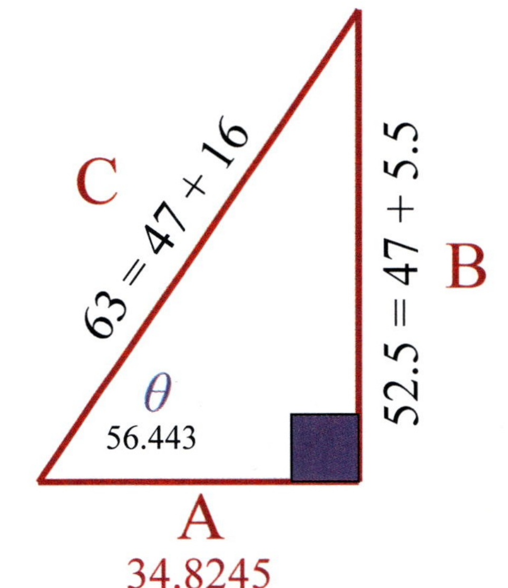

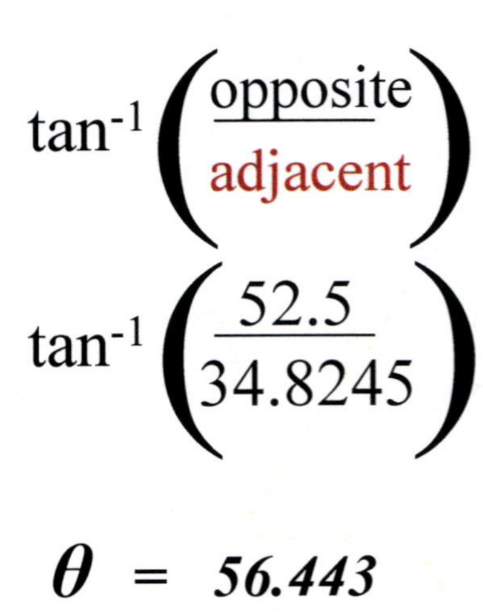

$$\tan^{-1}\left(\frac{\text{opposite}}{\text{adjacent}}\right)$$

$$\tan^{-1}\left(\frac{52.5}{34.8245}\right)$$

$$\theta = \mathbf{56.443}$$

    Advanced CNC Mill Programming and Applied Mathematics Level 2

Steps (cont):

G.   Draw small triangle to solve tangent point for upcoming IJ values.

H.   Using the previously calculated angle, solve the leg lengths of the small triangle.

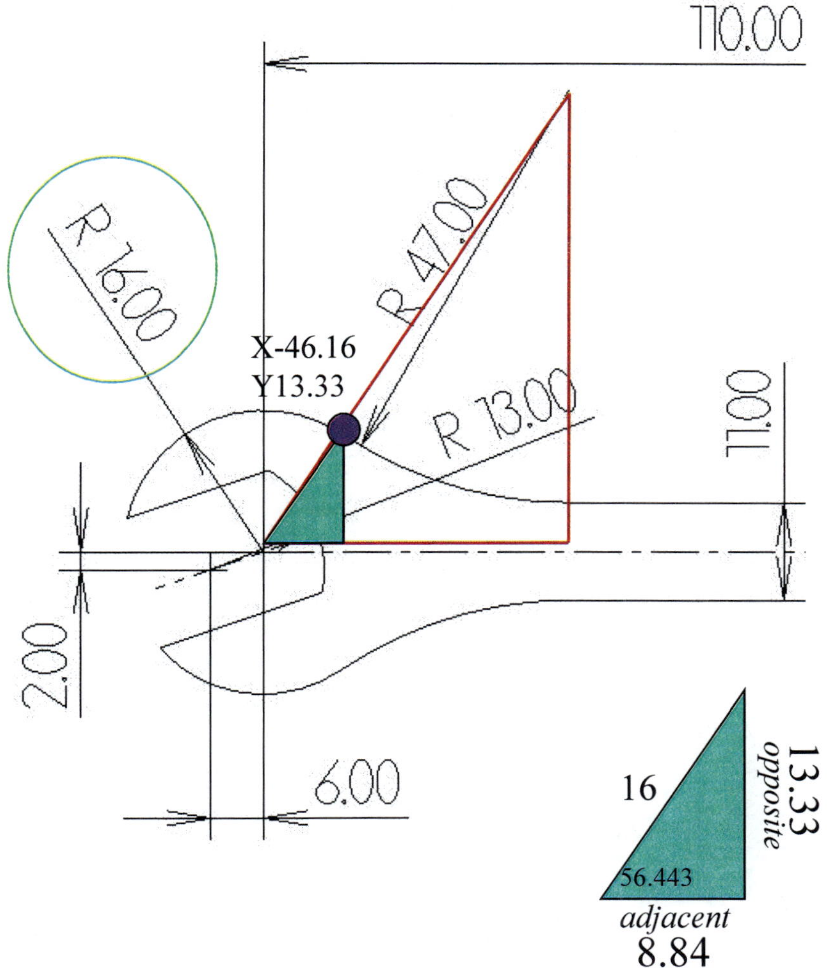

$$\sin\theta = \frac{opposite}{hypotenuse}$$

$$\sin(56.443) = \frac{opposite}{16}$$

$$0.8333 = \frac{opposite}{16}$$

$$opposite = 0.8333 \times 16$$

$$opposite = \mathbf{13.333}$$

$$\cos\theta = \frac{adjacent}{hypotenuse}$$

$$\cos(56.443) = \frac{adjacent}{16}$$

$$0.5528 = \frac{adjacent}{16}$$

$$adjacent = 0.5528 \times 16$$

$$adjacent = \mathbf{8.844}$$

   Advanced CNC Mill Programming and Applied Mathematics Level 2

Steps (cont):

I. Flip triangle upside down for use on the bottom half of the wrench.

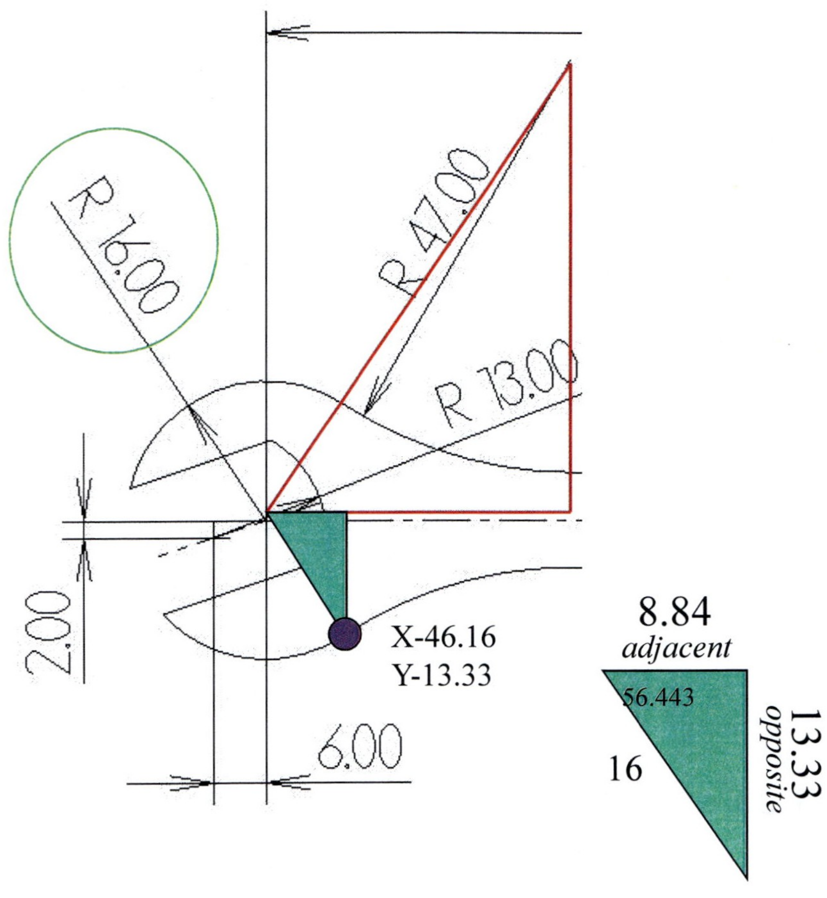

    Advanced CNC Mill Programming and Applied Mathematics Level 2

J. Calculate the sides of the triangle for IJ circular interpolation.

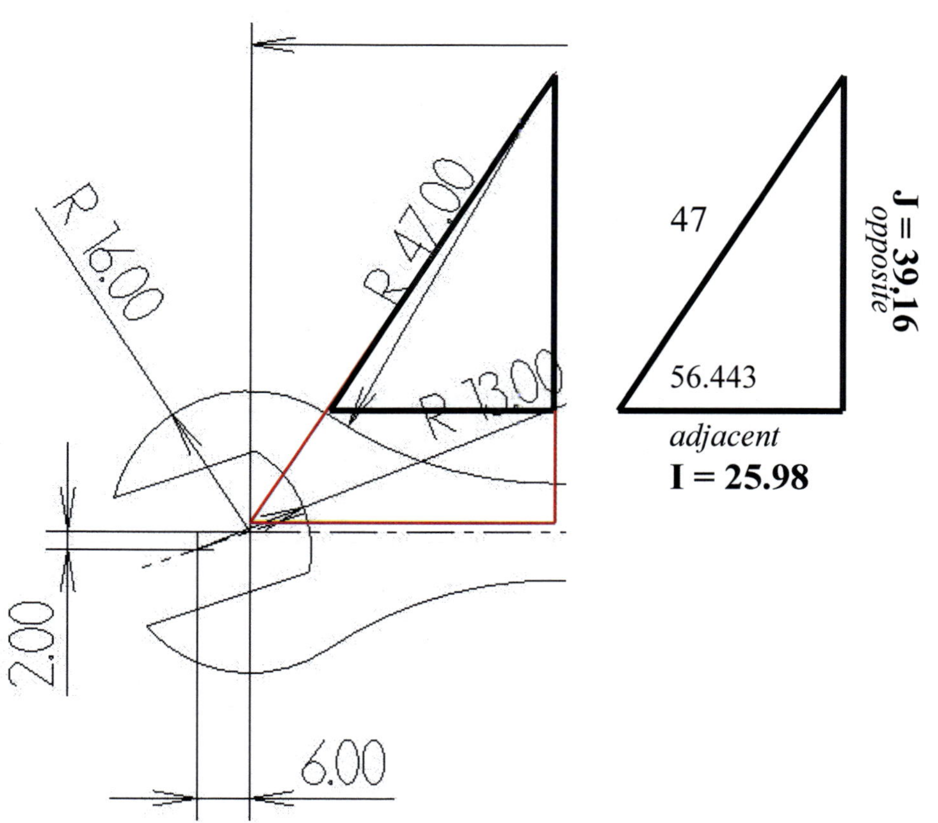

$$sin\ \theta = \frac{opposite}{hypotenuse}$$

$$sin(56.443) = \frac{opposite}{47}$$

$$0.8333 = \frac{opposite}{47}$$

$$opposite = 0.8333 \times 47$$

$$opposite = \textbf{39.16}$$

$$cos\ \theta = \frac{adjacent}{hypotenuse}$$

$$cos(56.443) = \frac{adjacent}{47}$$

$$0.5528 = \frac{adjacent}{47}$$

$$adjacent = 0.5528 \times 47$$

$$adjacent = \textbf{25.98}$$

     Advanced CNC Mill Programming and Applied Mathematics Level 2

Repeat steps to solve triangles on the right end of wrench.

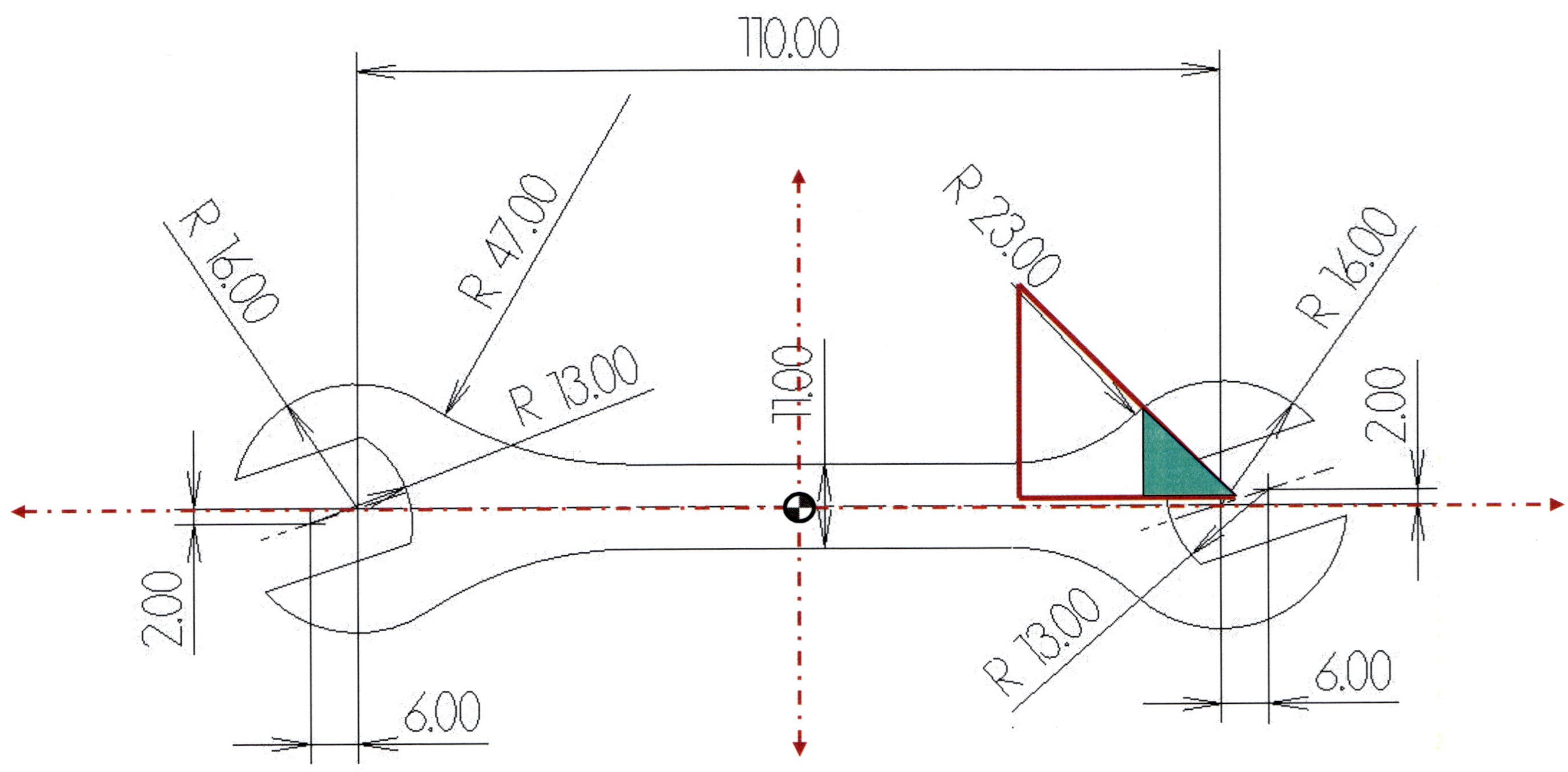

     Advanced CNC Mill Programming and Applied Mathematics Level 2

$A^2 = C^2 - B^2$

$A^2 = 39^2 - 28.5^2$

$A^2 = 1521 - 812.25$

$A = \sqrt{708.75}$

$A = \mathbf{26.622}$

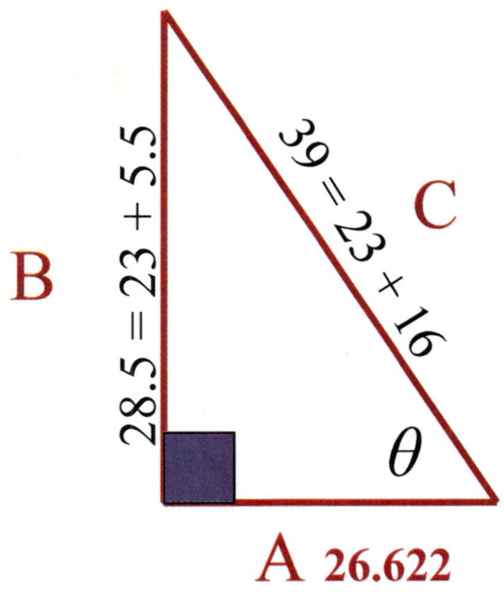

B

$28.5 = 23 + 5.5$

$39 = 23 + 16$

C

θ

$A\ \mathbf{26.622}$

$\tan^{-1}\left(\dfrac{\text{opposite}}{\text{adjacent}}\right)$

$\tan^{-1}\left(\dfrac{28.5}{26.622}\right)$

$\tan^{-1}\ 1.0705$

$\theta = \mathbf{46.951^{o}}$

     Advanced CNC Mill Programming and Applied Mathematics Level 2

$$sin\ \theta = \frac{opposite}{hypotenuse}$$

$$sin(46.951) = \frac{opposite}{16}$$

$$0.73077 = \frac{opposite}{16}$$

$$opposite = 0.73077 \times 16$$

$$opposite = \mathbf{11.692}$$

$$cos\ \theta = \frac{adjacent}{hypotenuse}$$

$$cos(46.951) = \frac{adjacent}{16}$$

$$0.6826 = \frac{adjacent}{16}$$

$$adjacent = 0.6826 \times 16$$

$$adjacent = \mathbf{10.921}$$

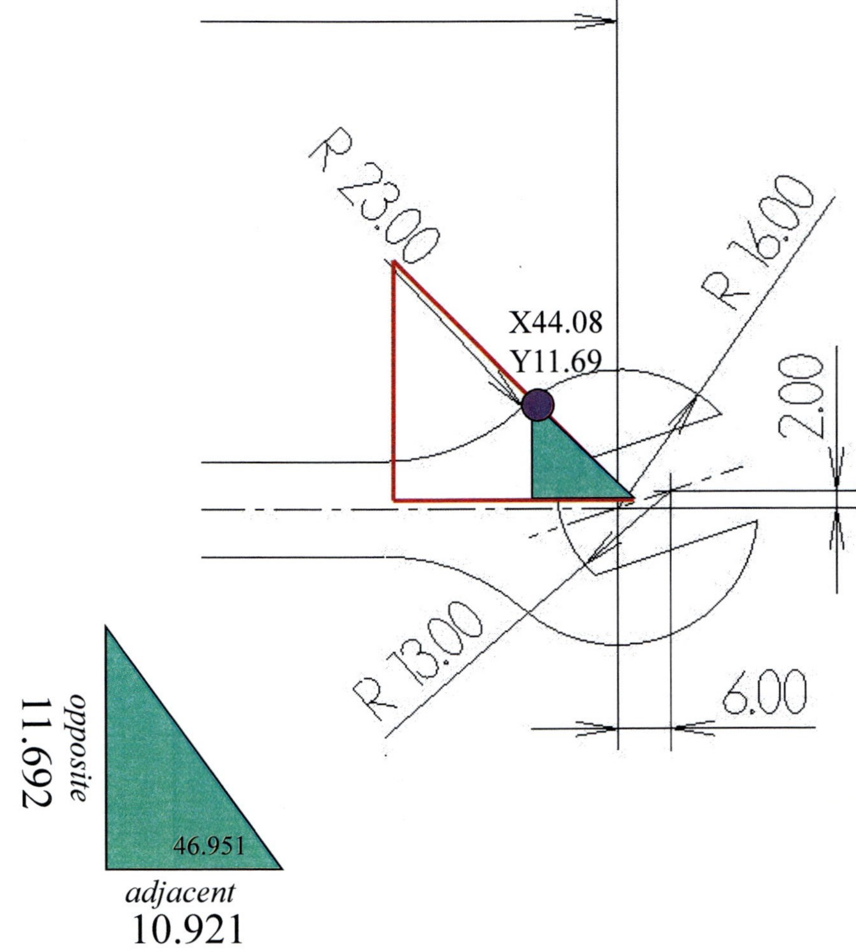

Advanced CNC Mill Programming and Applied Mathematics Level 2

Steps (cont):

Flip triangle upside down for use on the bottom half of the wrench.

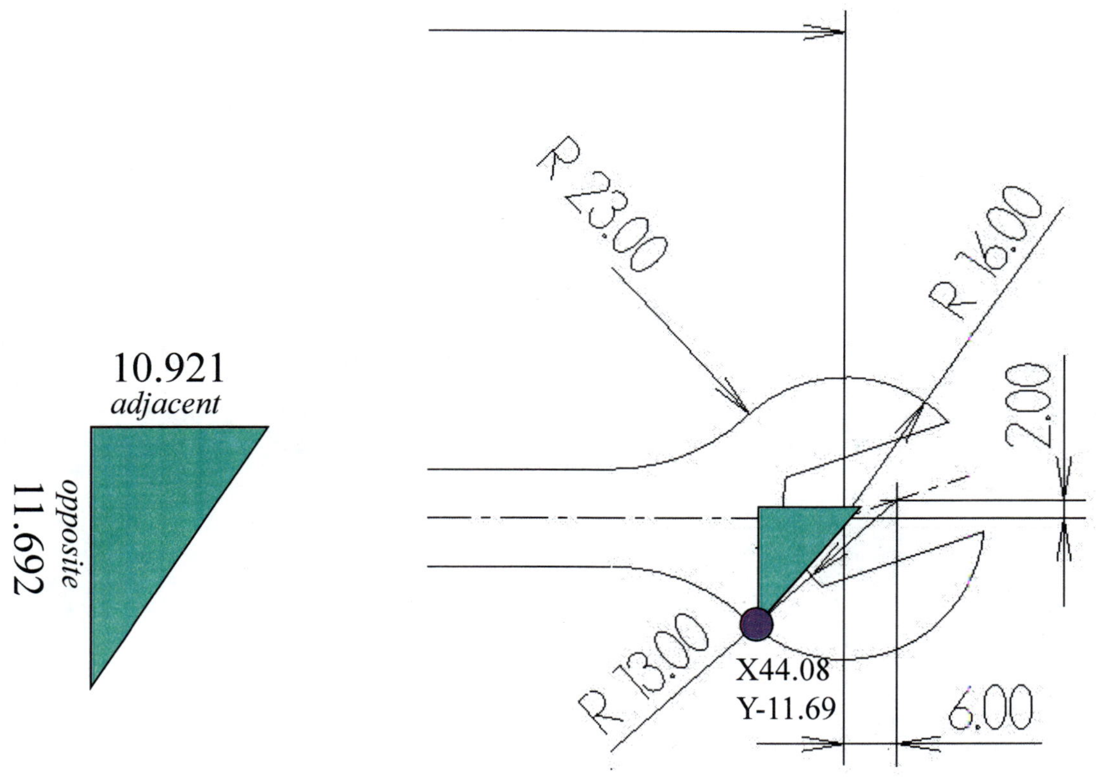

    Advanced CNC Mill Programming and Applied Mathematics Level 2

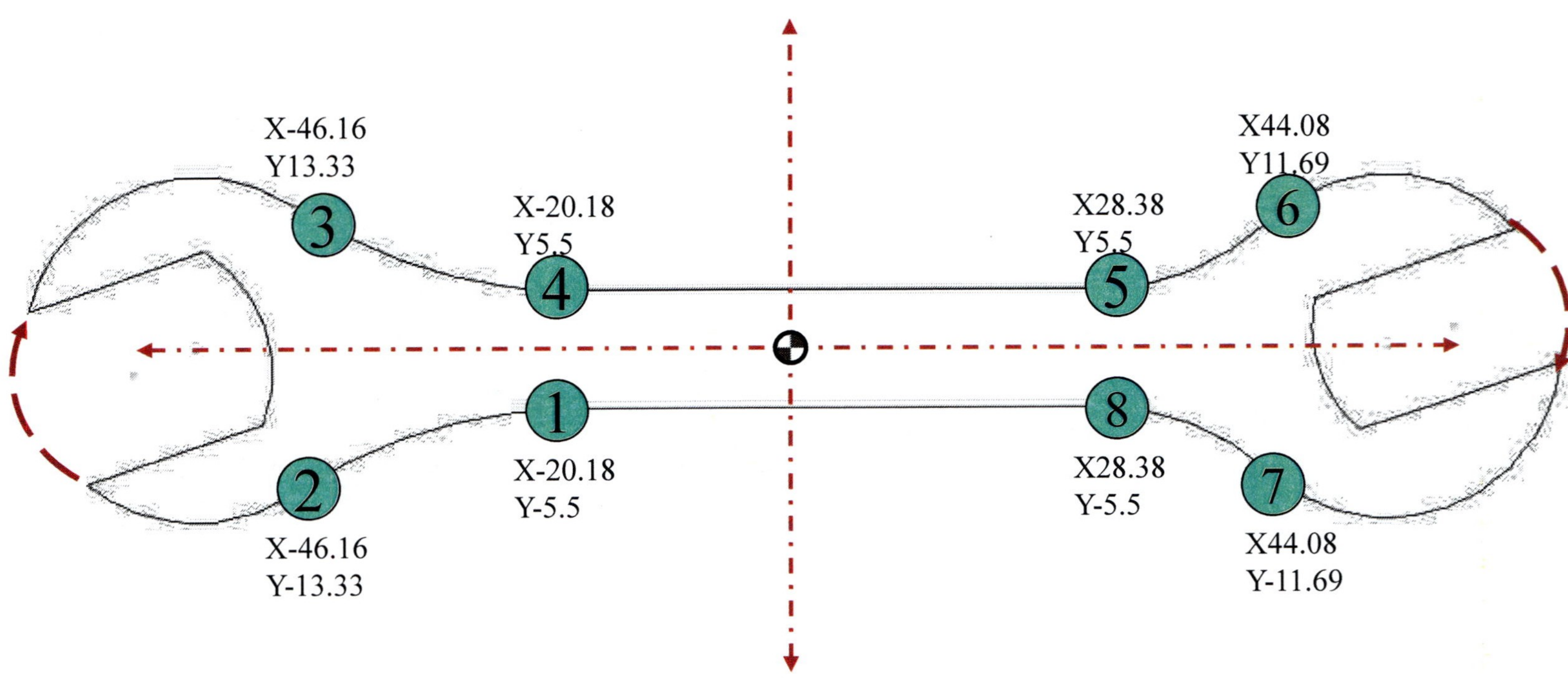

Copyright © Immersive Engineering 000112708I2   Advanced CNC Mill Programming and Applied Mathematics Level 2

Mill Project 404-1: Outer Profile

```
%
O4041  ( ORIGIN CENTER)
(Outer Profile of wrench – Without slots)
G28 G91 G0 Z0
G54 G90 G0 X0 Y-35.5 (Start of 15mm diamond)
T6 M6 (3/8 END MILL)
S3000 M3
G43 H6 Z5. M8
G1 Z-4. F100.
G41 D6 X15.0 Y-20.5              (1ST BASE)
G3 X0 Y-5.5 R15.                 (2ND BASE)
G1 X-20.18                       (PT 1)
G3 X-46.16 Y-13.33 I0 J-47.0     (PT 2)
G2 X-46.16 Y13.33 I-8.84 J13.33  (PT 3)
G3 X-20.18 Y5.5 I25.98 J39.16    (PT 4)
G1 X28.38 Y5.5                   (PT 5)
G3 X44.08 Y11.69 R23.0           (PT 6)
G2 X44.08 Y-11.69 R-16.          (PT 7 ARC > 180 R-)
G3 X28.38 Y-5.5 R23.0            (PT 8)
G1 X0 Y-5.5                      (2nd BASE)
G3 X-15.0 Y-20.5 R15.0           (3rd BASE)
G1 G40 X0 Y-35.5                 (Home Base)
G28 G91 G0 Z0
M30
%
```

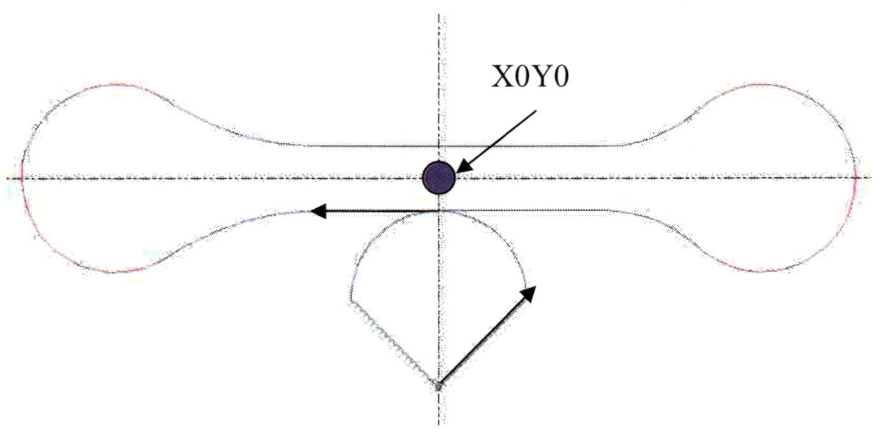

Mill Project 404-2: Slots with R13

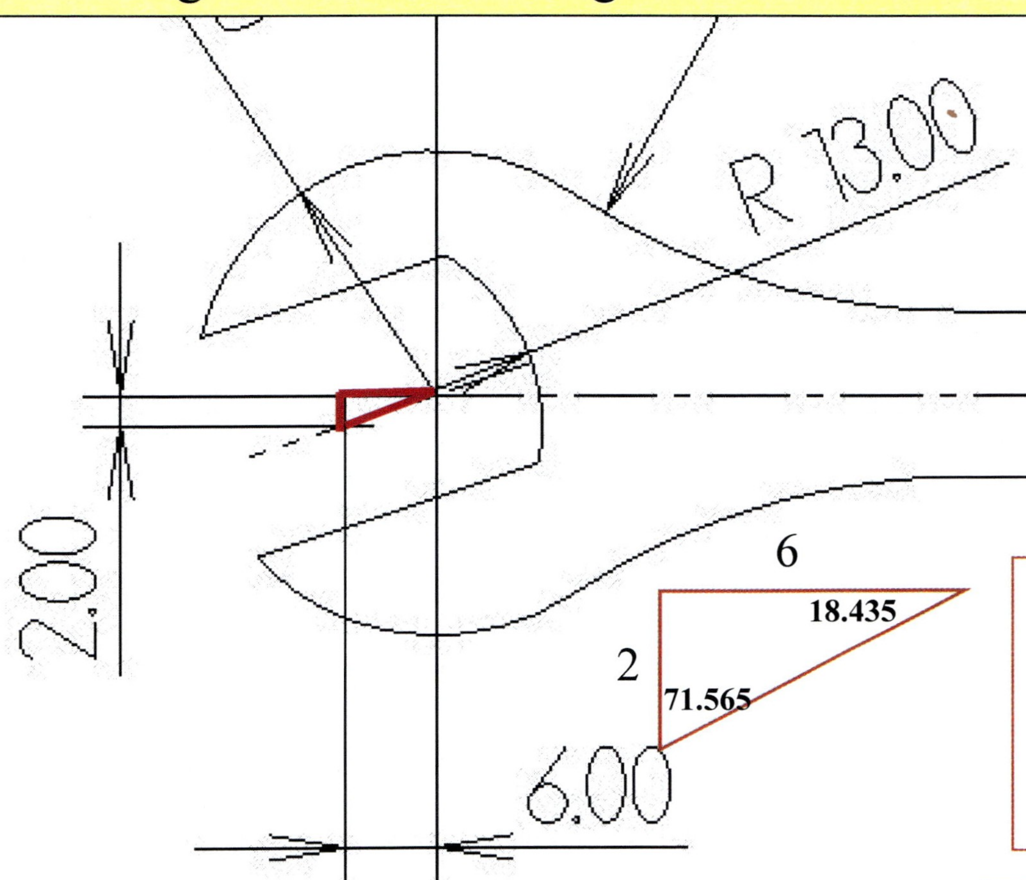

Rotation Angle

$90 - 71.565 = \textbf{18.435}°$

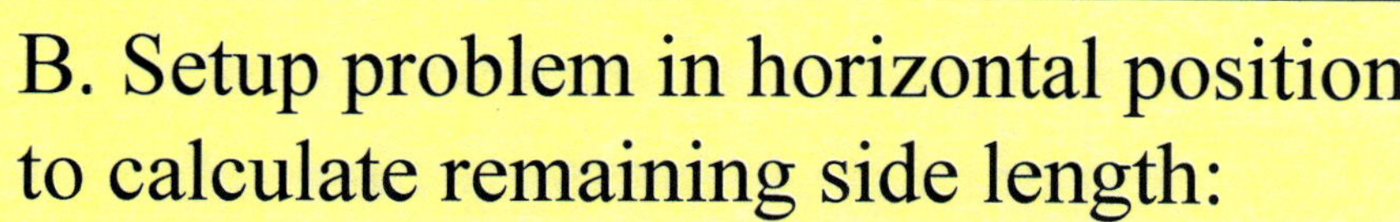

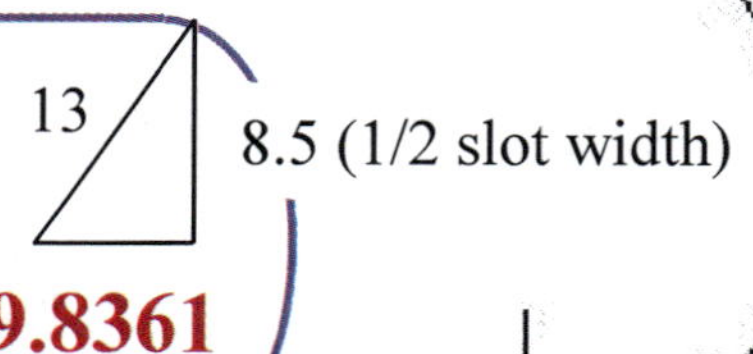

$$A^2 = C^2 - B^2$$

$$A^2 = 13^2 - 8.5^2$$

$$A^2 = 169 - 72.25$$

$$A = \sqrt{96.75}$$

$$A = 9.8361$$

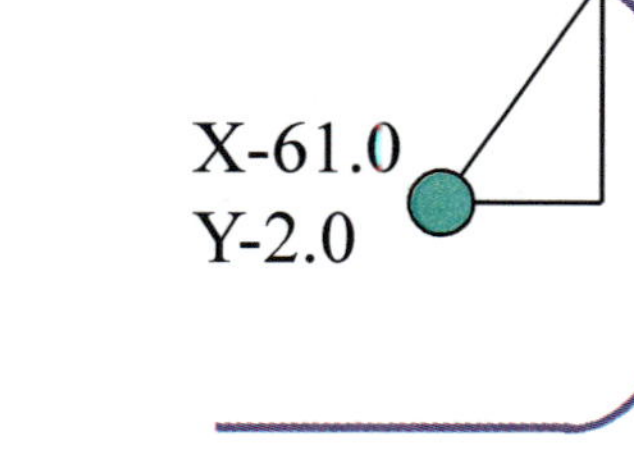

   Advanced CNC Mill Programming and Applied Mathematics Level 2

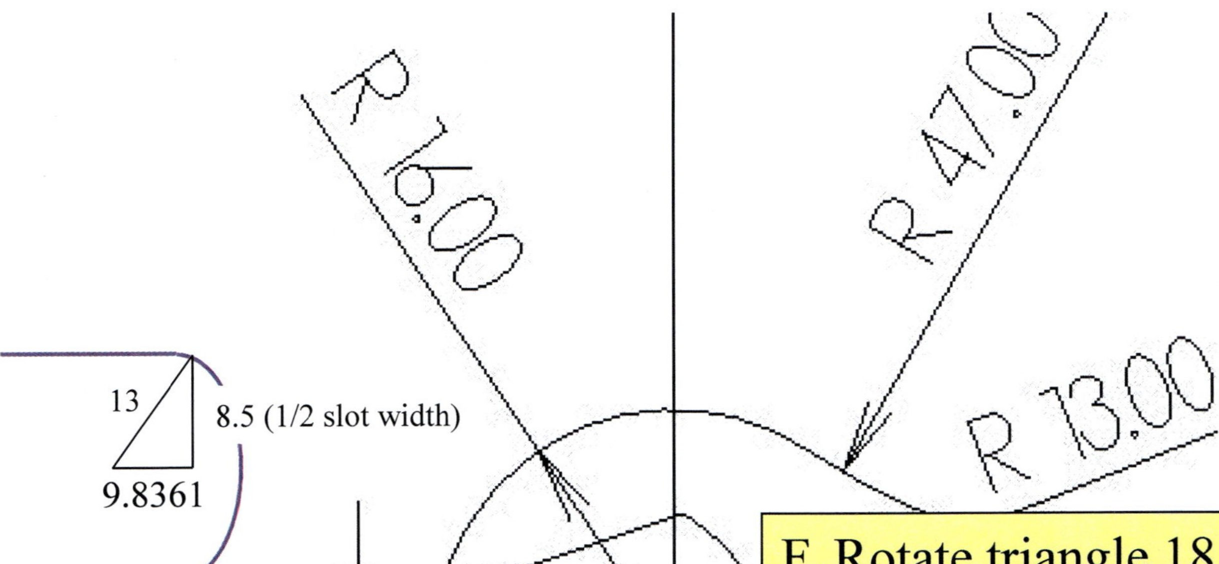

D. Calculate angle:

$$\tan^{-1} \frac{opposite}{adjacent}$$

$$\tan^{-1} \frac{8.5}{9.8361}$$

$$\theta = \mathbf{40.832°}$$

E. Add Rotation:

$$\begin{array}{r} 40.832 \\ +\ 18.435 \\ \hline \mathbf{59.267°} \end{array}$$

F. Rotate triangle 18.435° and calculate using Polar coordinates (R theta).

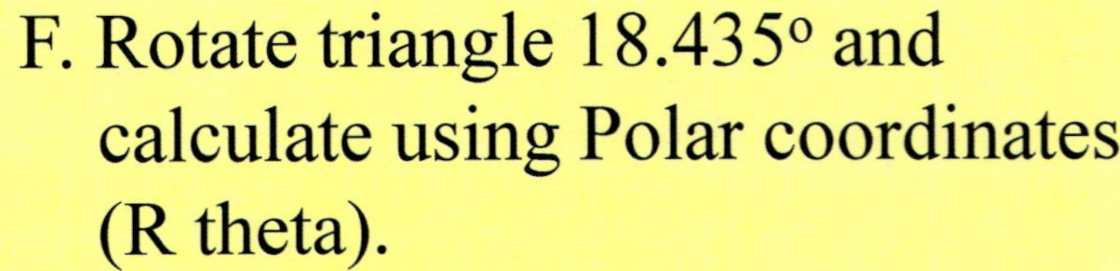

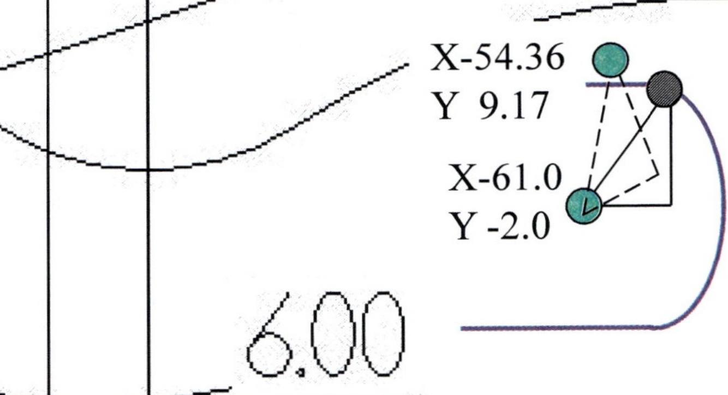

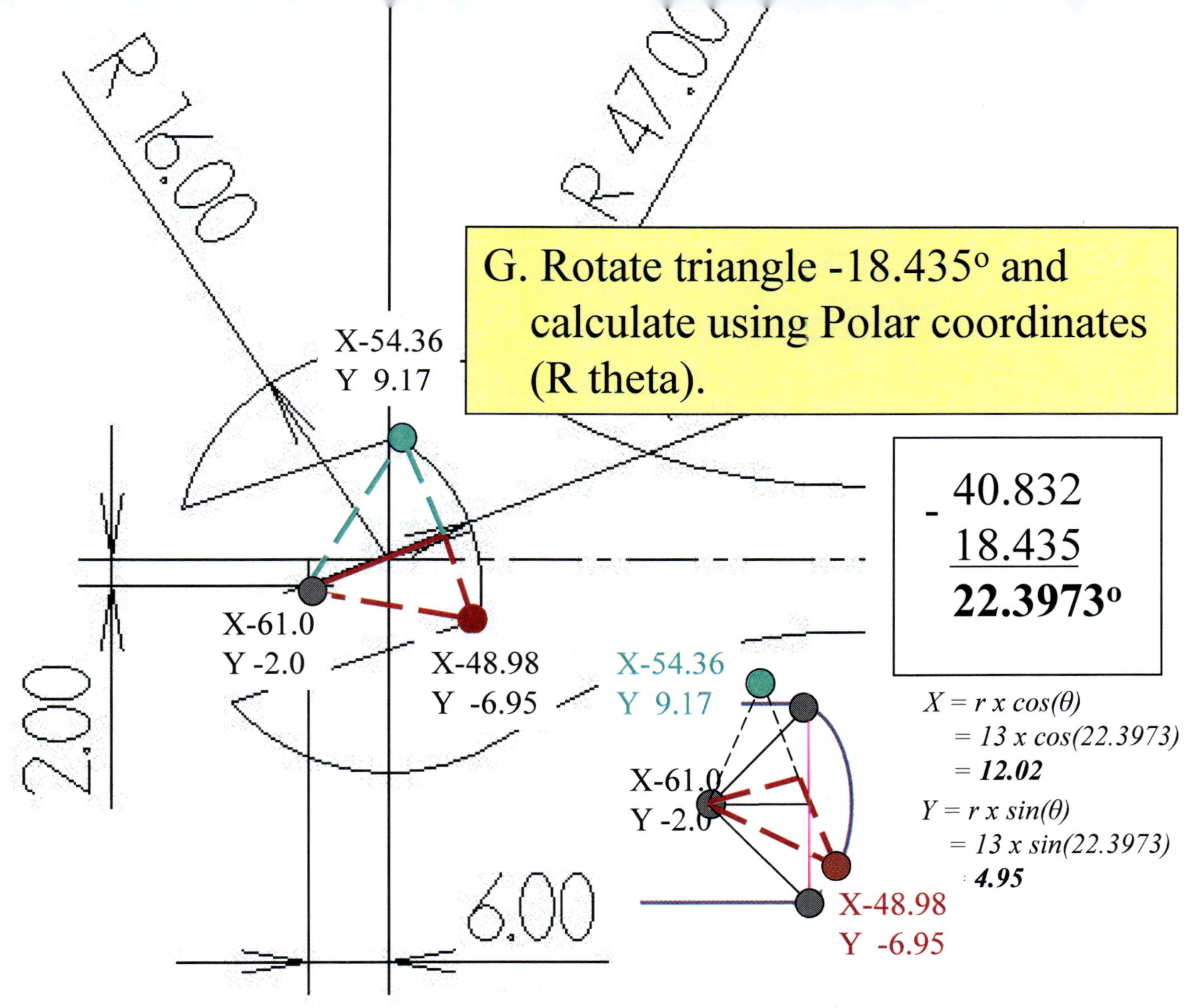

$$-\begin{array}{r} 40.832 \\ 18.435 \\ \hline \mathbf{22.3973º} \end{array}$$

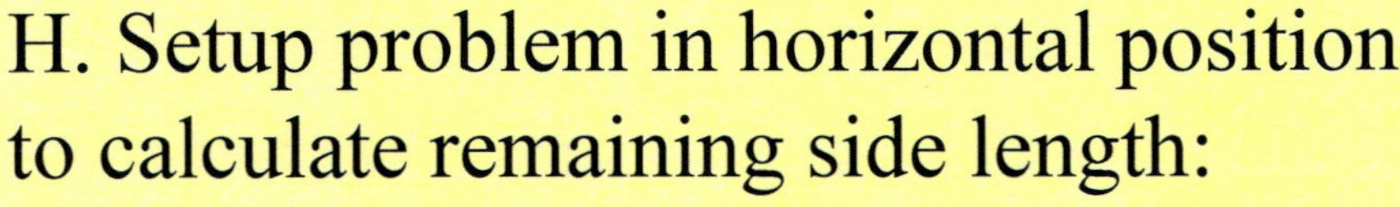

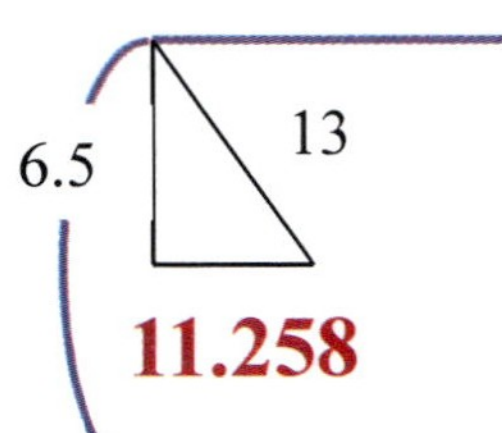

11.00

2.00

6.5   13

11.258

$$A^2 = C^2 - B^2$$

$$A^2 = 13^2 - 6.5^2$$

$$A^2 = 169 - 42.25$$

$$A = \sqrt{126.75}$$

$$A = \mathbf{11.258}$$

6.00

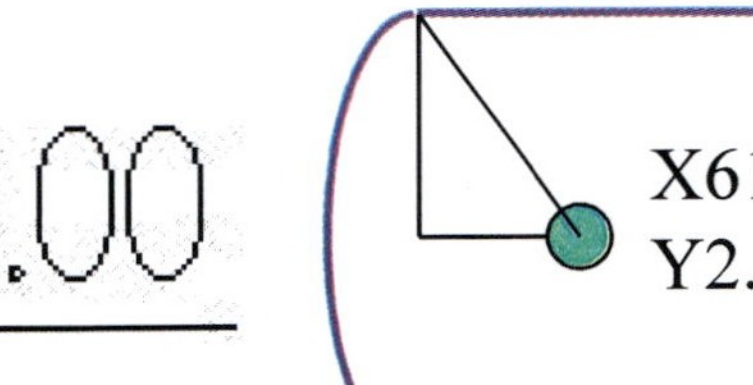

150

    Advanced CNC Mill Programming and Applied Mathematics Level 2

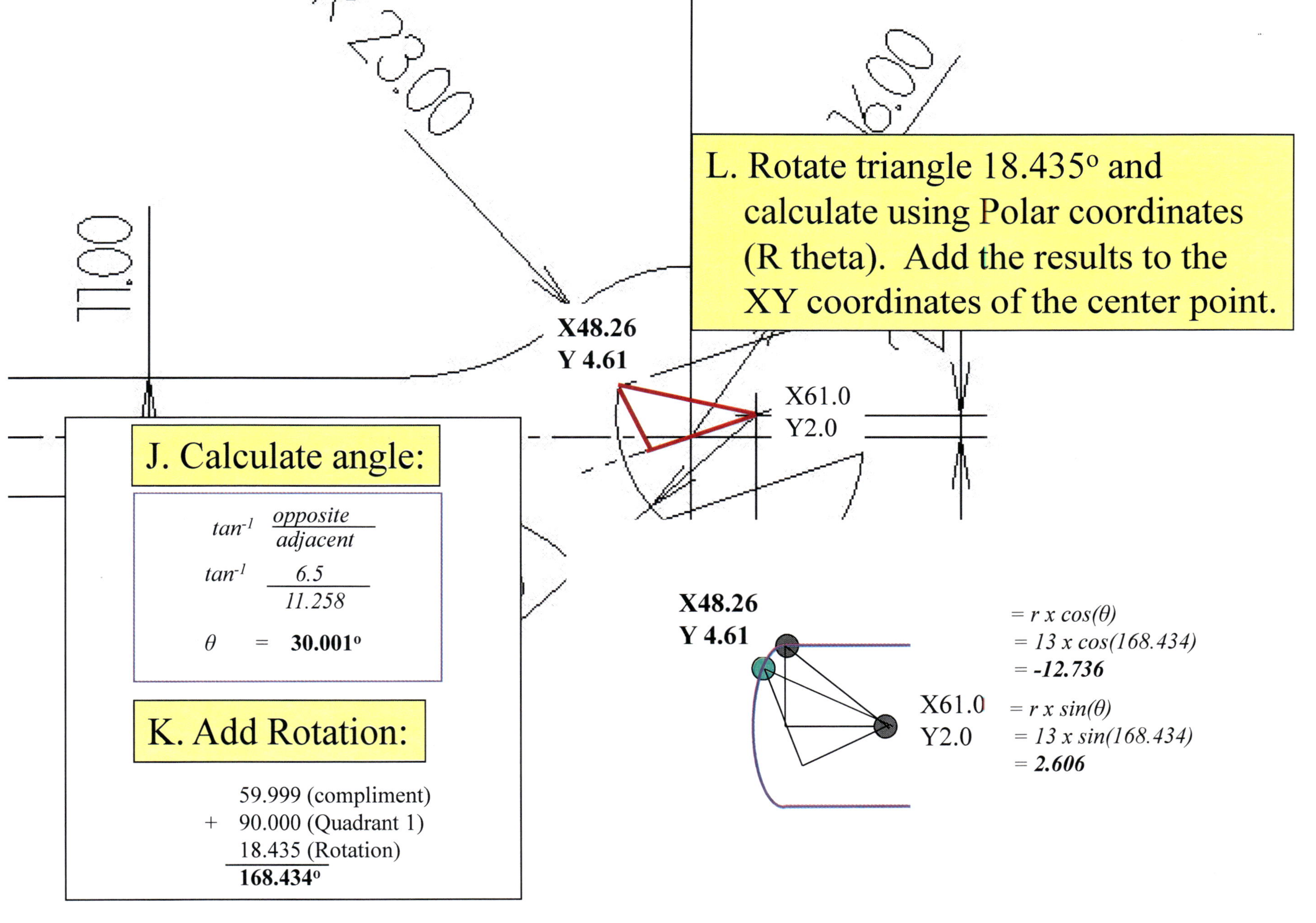
R 23.00
11.00
R 16.00
L. Rotate triangle 18.435º and calculate using Polar coordinates (R theta). Add the results to the XY coordinates of the center point.

X48.26
Y 4.61

X61.0
Y2.0

J. Calculate angle:

$tan^{-1} \dfrac{opposite}{adjacent}$

$tan^{-1} \dfrac{6.5}{11.258}$

θ = 30.001º

K. Add Rotation:

59.999 (compliment)
+ 90.000 (Quadrant 1)
18.435 (Rotation)
168.434º

X48.26
Y 4.61

X61.0
Y2.0

= r x cos(θ)
= 13 x cos(168.434)
= -12.736

= r x sin(θ)
= 13 x sin(168.434)
= 2.606

$$168.434$$
$$+\ 59.999$$
$$\mathbf{228.433°}$$

$X = r \times cos(\theta)$
  $= 13 \times cos(228.433)$
  $= \mathbf{-8.625}$

$Y = r \times sin(\theta)$
  $= 13 \times sin(228.433)$
  $= \mathbf{-9.726}$

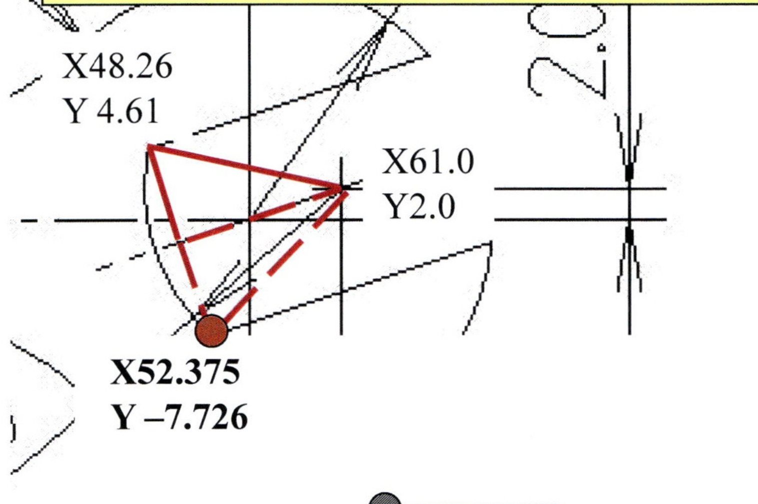

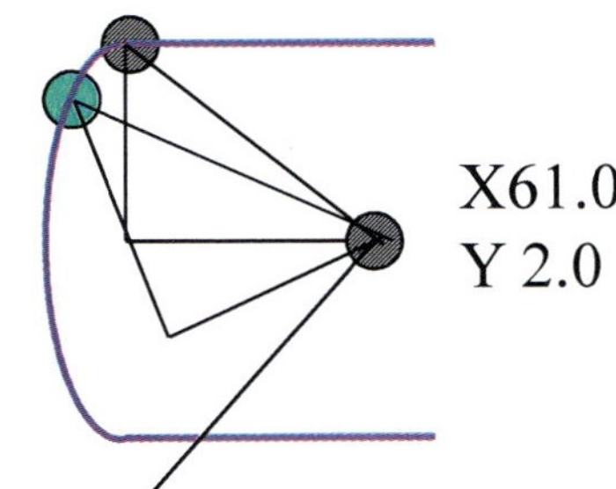

152

A. We'll use the same calculated angle for the rotational amount.

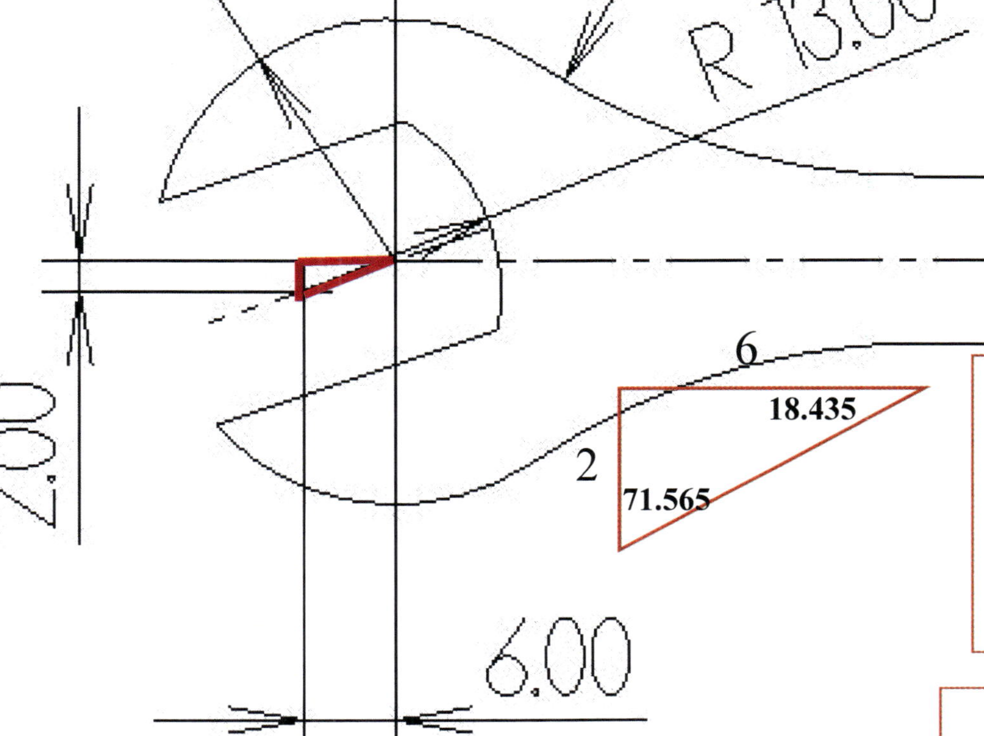

$$tan^{-1} \frac{opposite}{adjacent}$$

$$tan^{-1} \frac{6.0}{2.0}$$

$$\theta \quad = \quad 71.565$$

Rotation Angle

$$90 - 71.565 = \textbf{18.435}^\textbf{o}$$

153

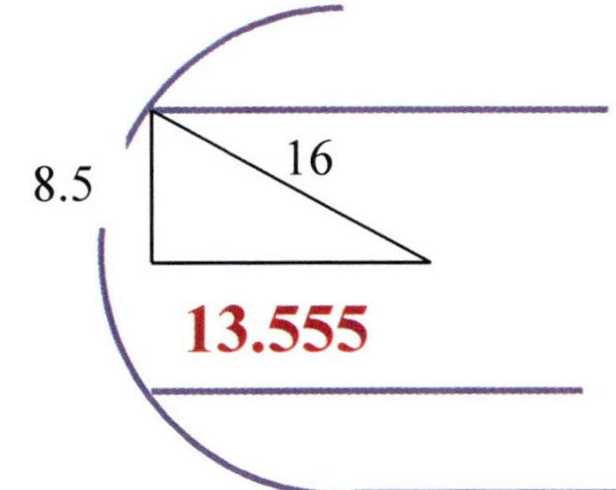

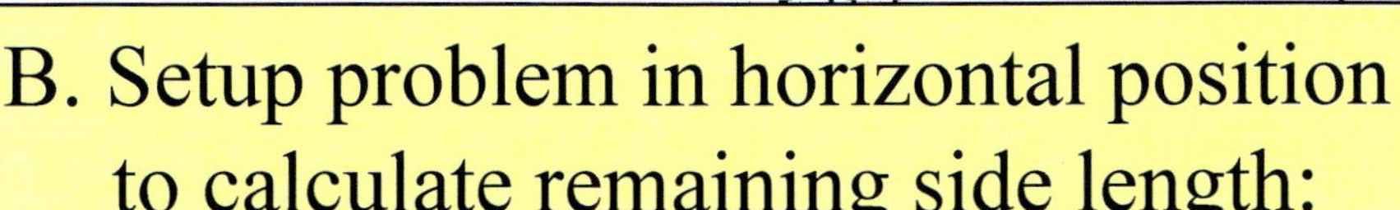

$$A^2 = C^2 - B^2$$

$$A^2 = 16^2 - 8.5^2$$

$$A^2 = 256 - 72.25$$

$$A = \sqrt{183.75}$$

$$A = 13.555$$

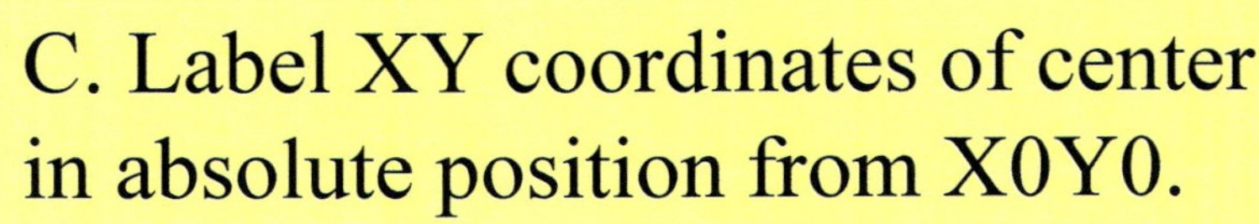

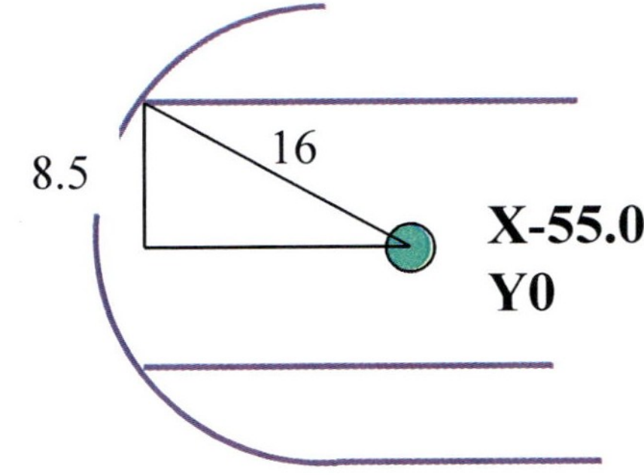

154

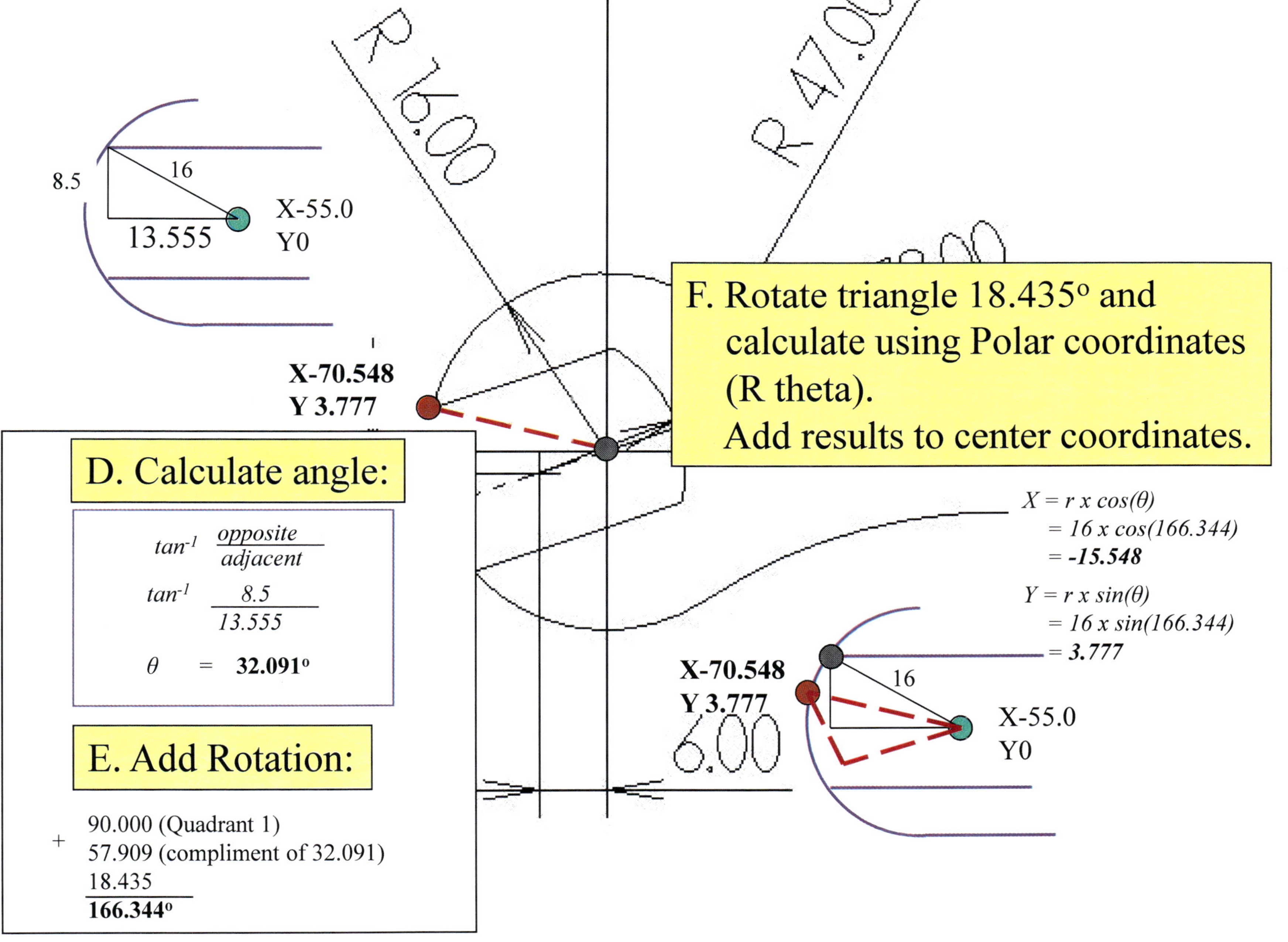

Advanced CNC Mill Programming and Applied Mathematics Level 2

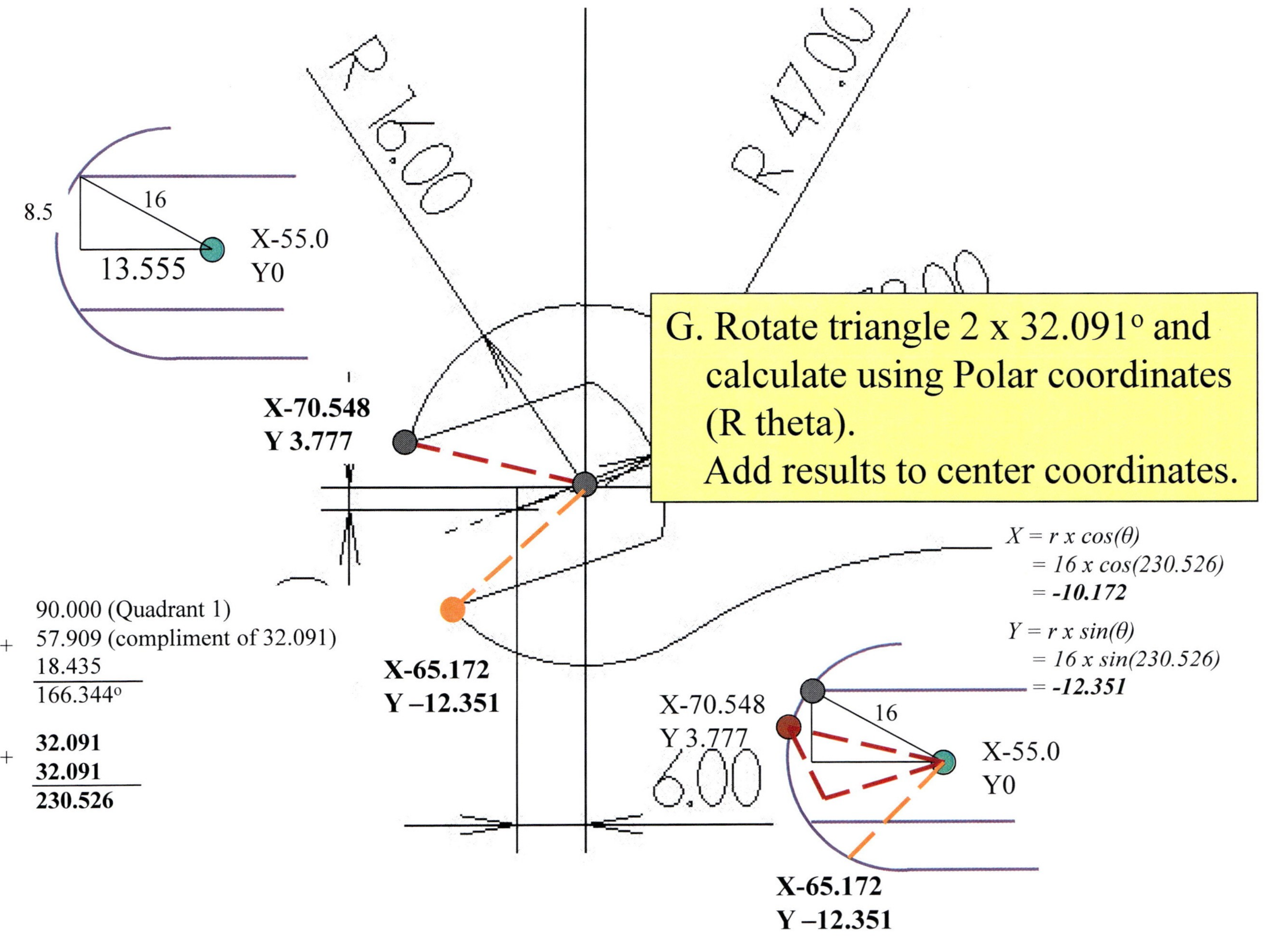

Advanced CNC Mill Programming and Applied Mathematics Level 2

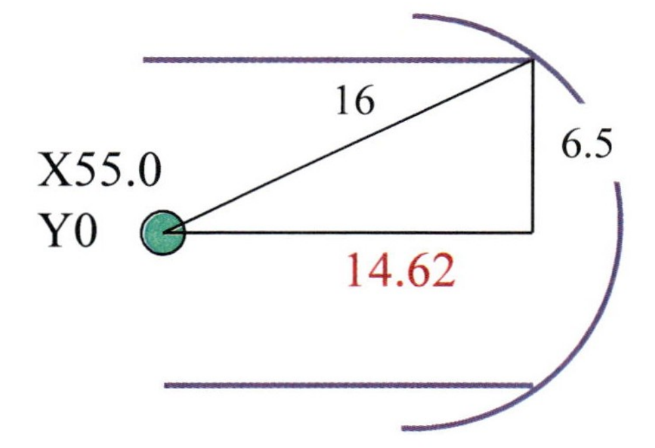

H. Setup problem in horizontal position to calculate remaining side length:

I. Label center coordinates.

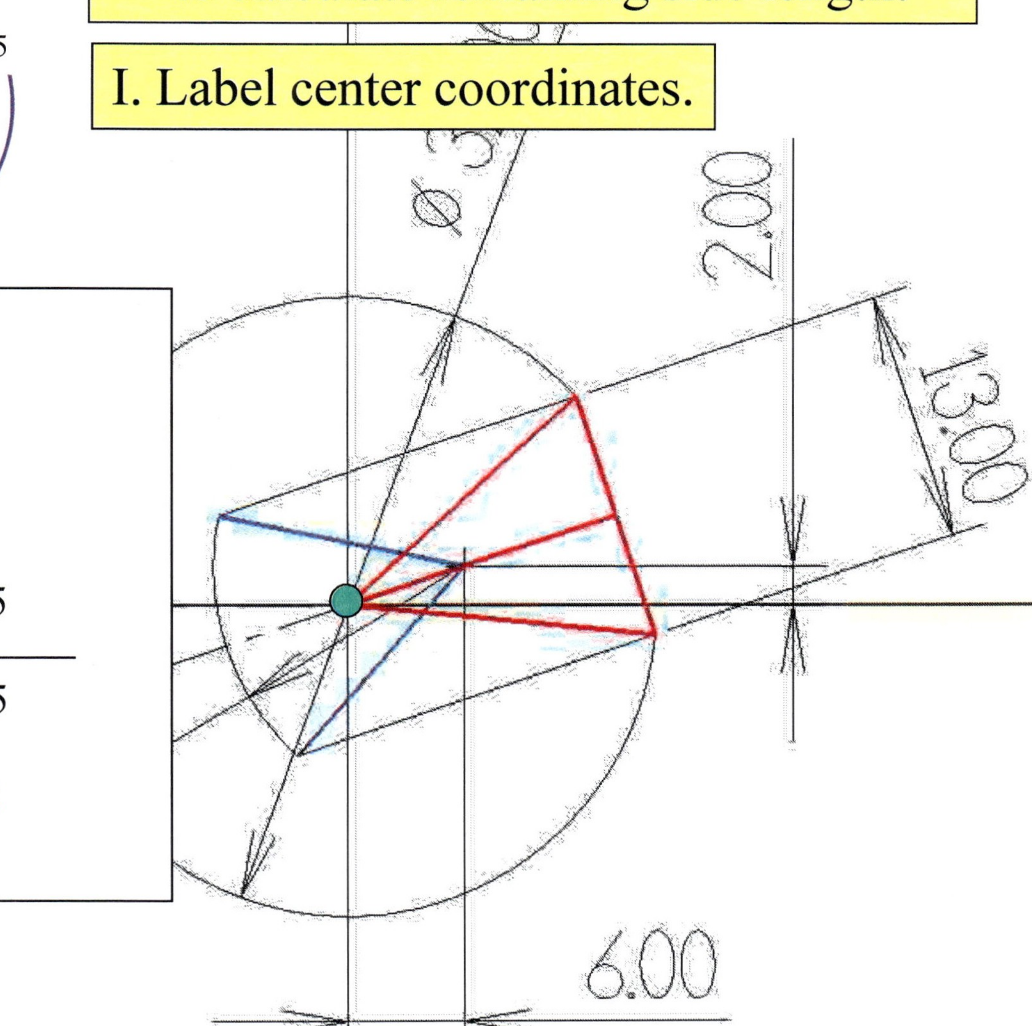

   Advanced CNC Mill Programming and Applied Mathematics Level 2

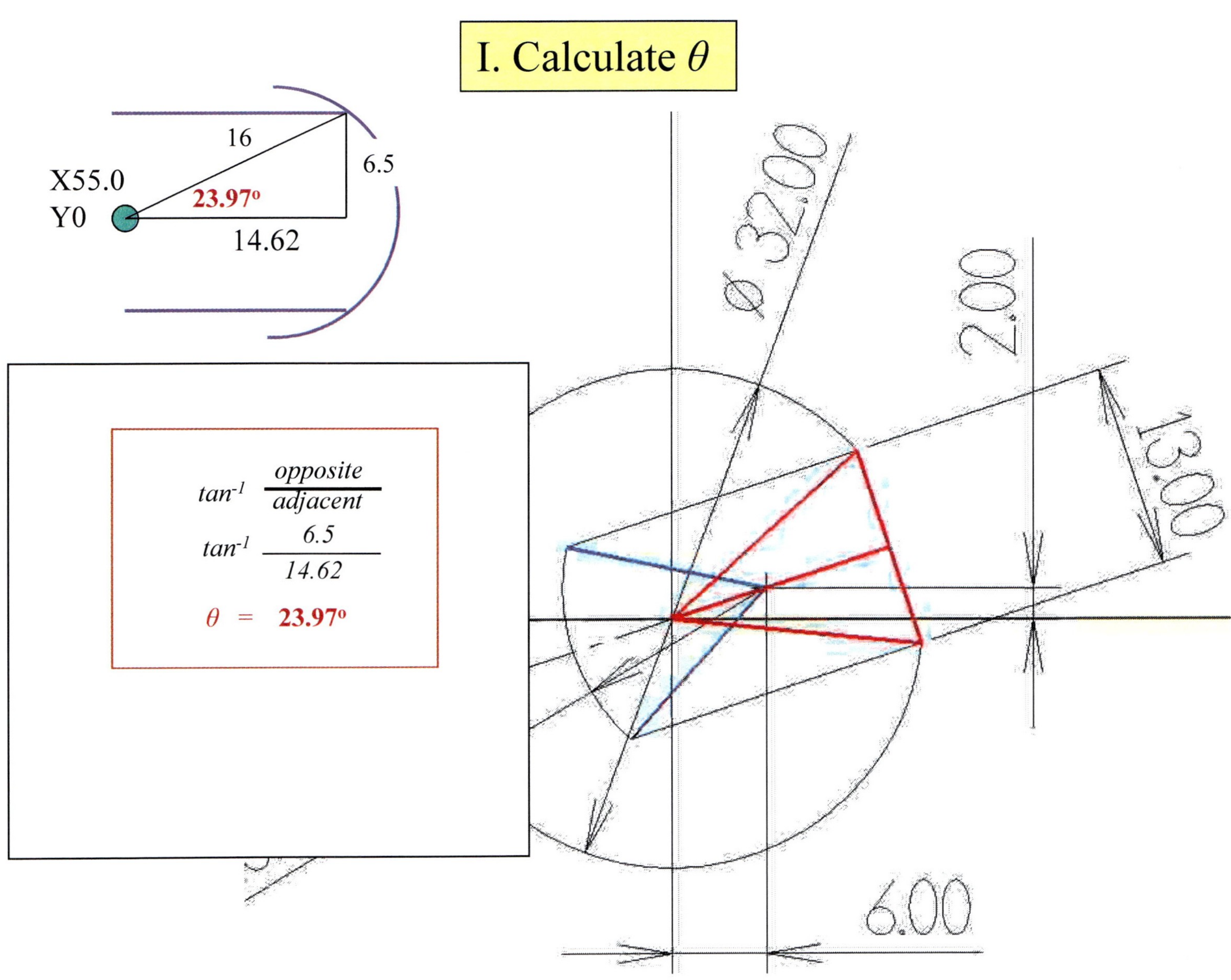

16
6.5
X55.0
Y0
23.97°
14.62
opposite
adjacent
tan^{-1}
tan^{-1} $\dfrac{6.5}{14.62}$
$\theta = 23.97°$
Ø 32.00
2.00
13.00
6.00

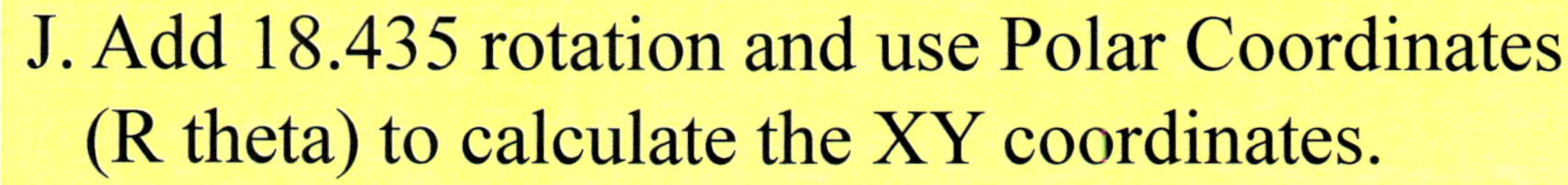

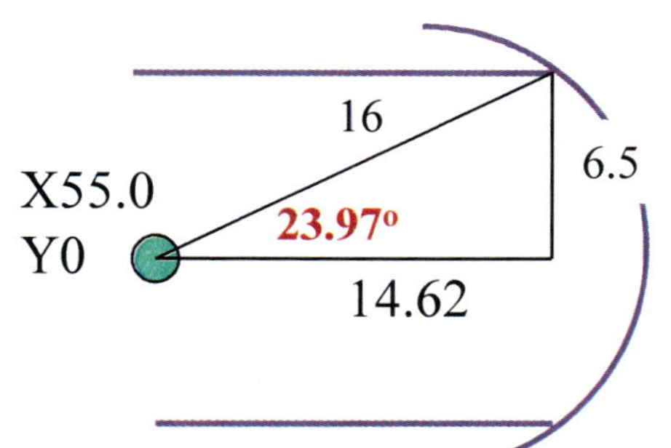

$X = r \times \cos(\theta)$
  $= 16 \times \cos(42.405)$
  $= 11.814$

$Y = r \times \sin(\theta)$
  $= 16 \times \sin(42.405)$
  $= 10.79$

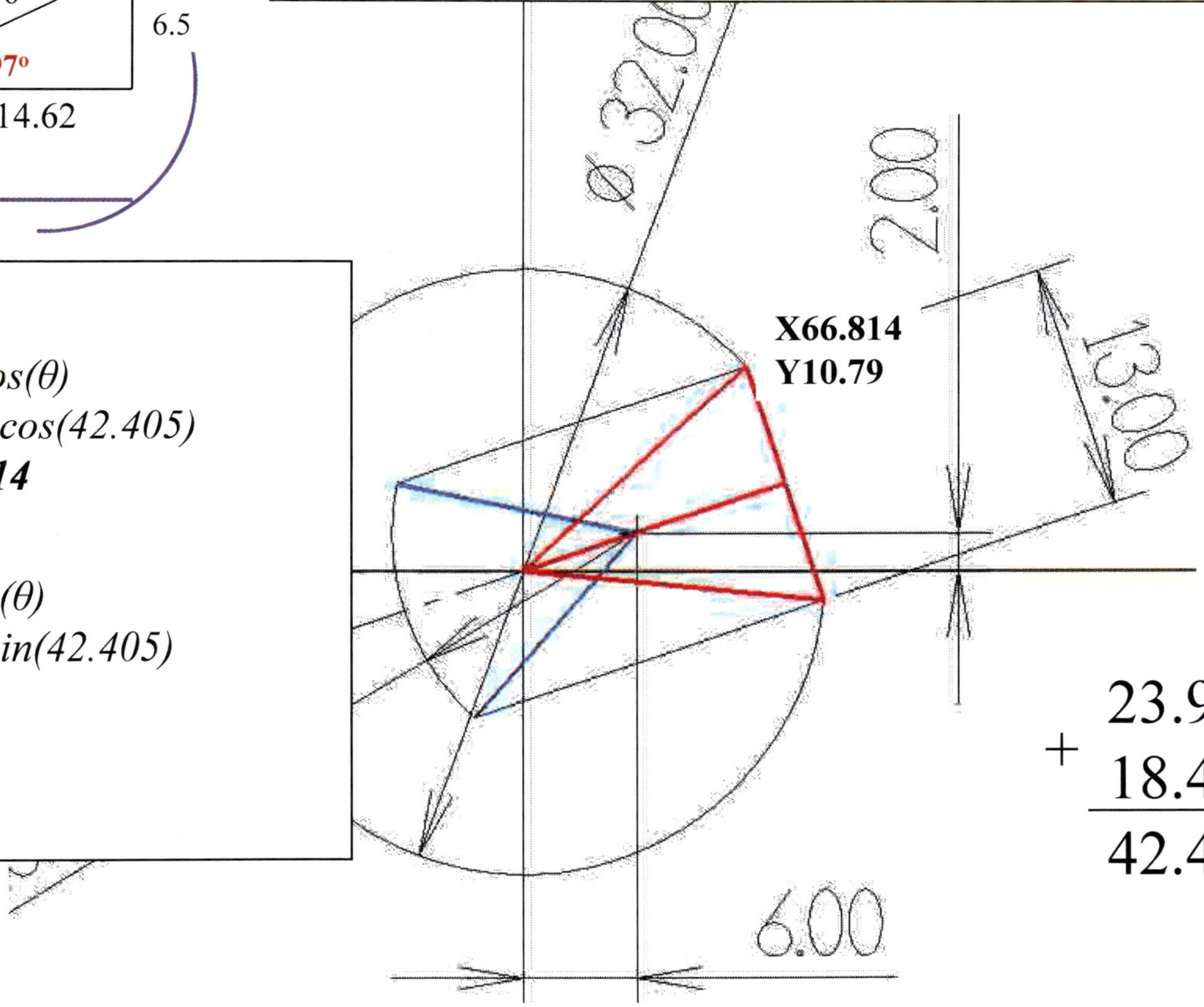

X66.814
Y10.79

Ø 32.00

2.00

13.00

6.00

$$\begin{array}{r} 23.970 \\ +\ 18.435 \\ \hline 42.405 \end{array}$$

159

    Advanced CNC Mill Programming and Applied Mathematics Level 2

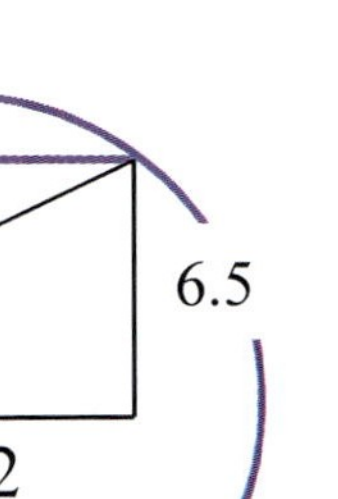

$X = r \times cos(\theta)$
$= 16 \times cos(-5.535)$
$= \mathbf{15.925}$

$Y = r \times sin(\theta)$
$= 16 \times sin(-5.535)$
$= \mathbf{-1.543}$

Advanced CNC Mill Programming and Applied Mathematics Level 2

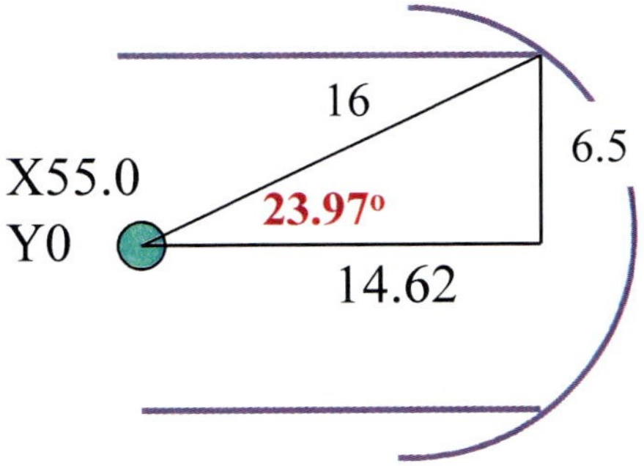

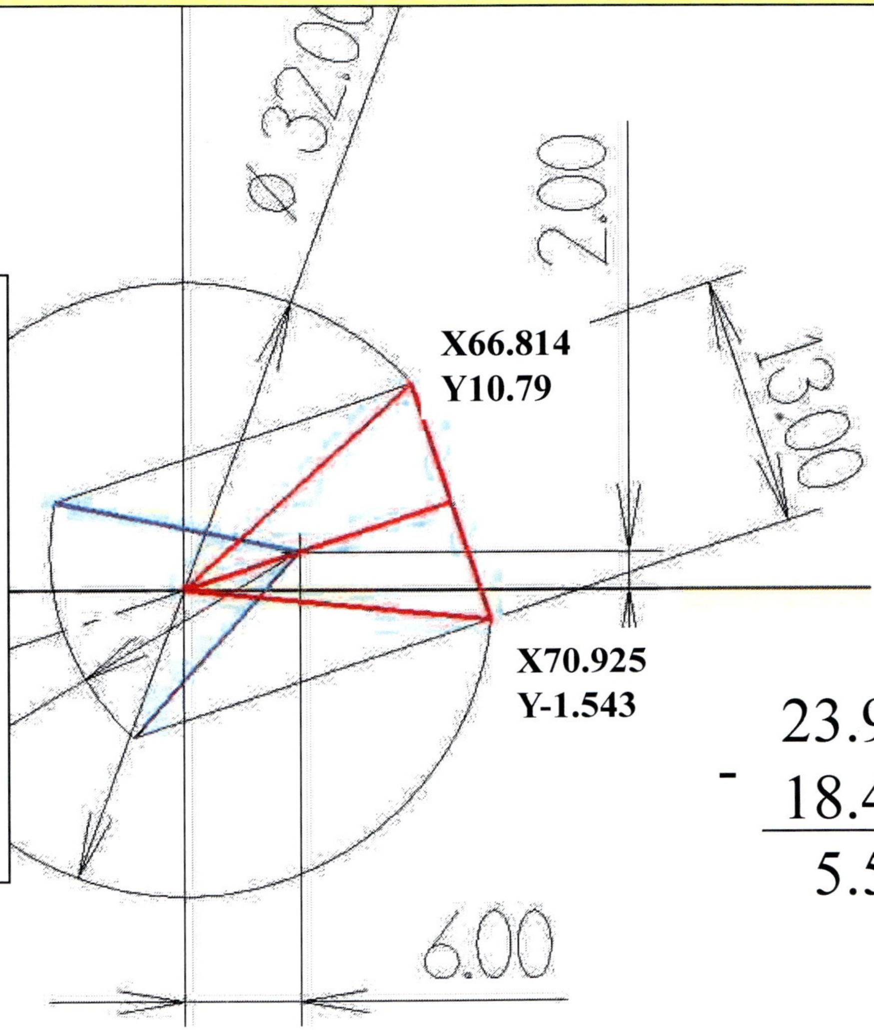

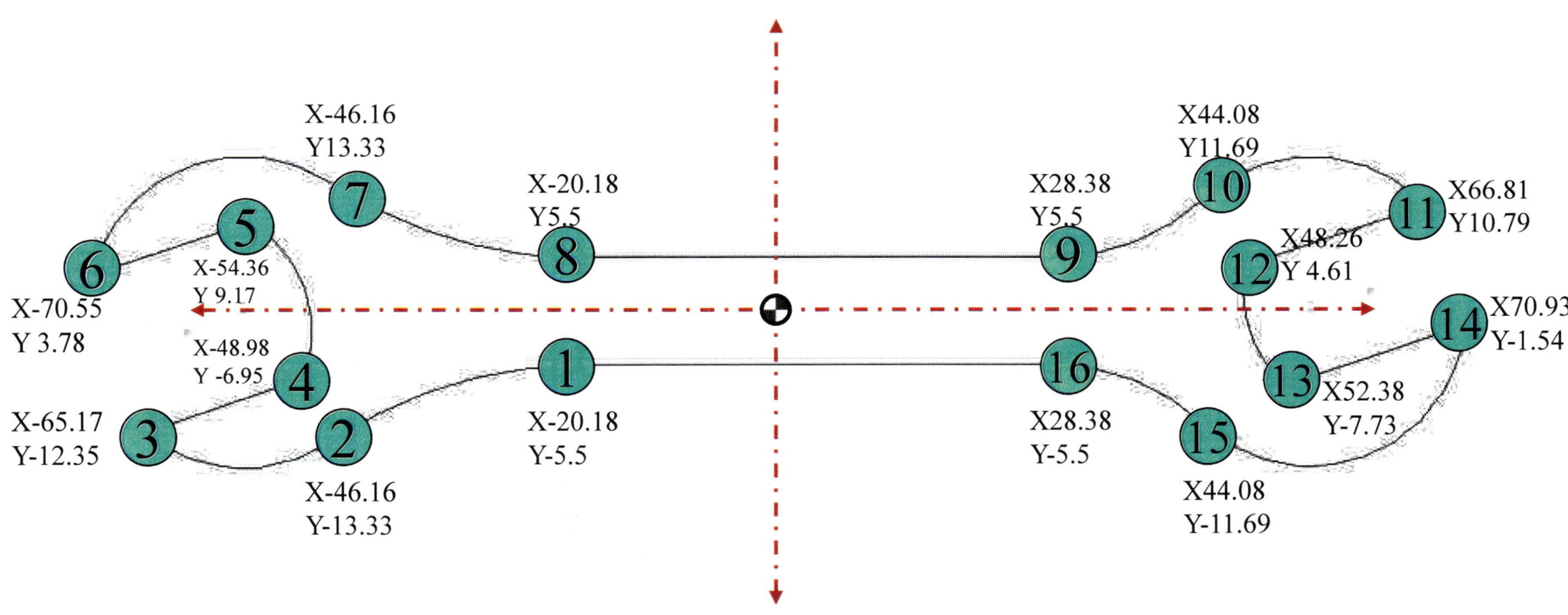

X-46.16
Y13.33
X-20.18
Y5.5
X28.38
Y5.5
X44.08
Y11.69
X66.81
Y10.79
X-54.36
Y 9.17
X48.26
Y 4.61
X-70.55
Y 3.78
X-48.98
Y -6.95
X70.93
Y-1.54
X-65.17
Y-12.35
X52.38
Y-7.73
X-20.18
Y-5.5
X28.38
Y-5.5
X-46.16
Y-13.33
X44.08
Y-11.69
1
2
3
4
5
6
7
8
9
10
11
12
13
14
15
16

```
%
O4042  ( ORIGIN CENTER)
(Profile wrench – including slots)
G28 G91 G0 Z0
G54 G90 G0 X0 Y-35.5                (Start of 15mm diamond)
T6 M6                               (3/8 END MILL)
S3000 M3
G43 H6 Z5. M8
G1 Z-4. F100.
G41 D6 X15.0 Y-20.5                 (1ST BASE)
G3 X0 Y-5.5 R15.                    (2ND BASE)
G1 X-20.18                          (PT 1)
G3 X-46.16 Y-13.33 I0 J-47.0        (PT 2)
G2 X-65.17 Y-12.35 R16.            (PT 3)
G1 X-48.98 Y-6.95                   (PT 4)
G3 X-54.36 Y9.17 R13.              (PT 5)
G1 X-70.55 Y3.78                    (PT 6)
G2 X-46.16 Y13.33 R16.            (PT 7)
G3 X-20.18 Y5.5 I25.98 J39.16     (PT 8)
G1 X28.38 Y5.5                      (PT 9)
G3 X44.08 Y11.69 R23.0            (PT 10)
G2 X66.81 Y10.79 R16.            (PT 11)
G1 X48.26 Y4.61                     (PT 12)
G3 X52.38 Y-7.73 R13.             (PT 13)
G1 X70.93 Y-1.54                    (PT 14)
G2 X44.08 Y-11.69 R16.           (PT 15)
G3 X28.38 Y-5.5 R23.0            (PT 16)
G1 X0 Y-5.5                         (2nd BASE)
G3 X-15.0 Y-20.5 R15.0            (3rd BASE)
G1 G40 X0 Y-35.5                    (Home Base)
G28 G91 G0 Z0
M30
%
```

     Advanced CNC Mill Programming and Applied Mathematics Level 2

Mill Project 405

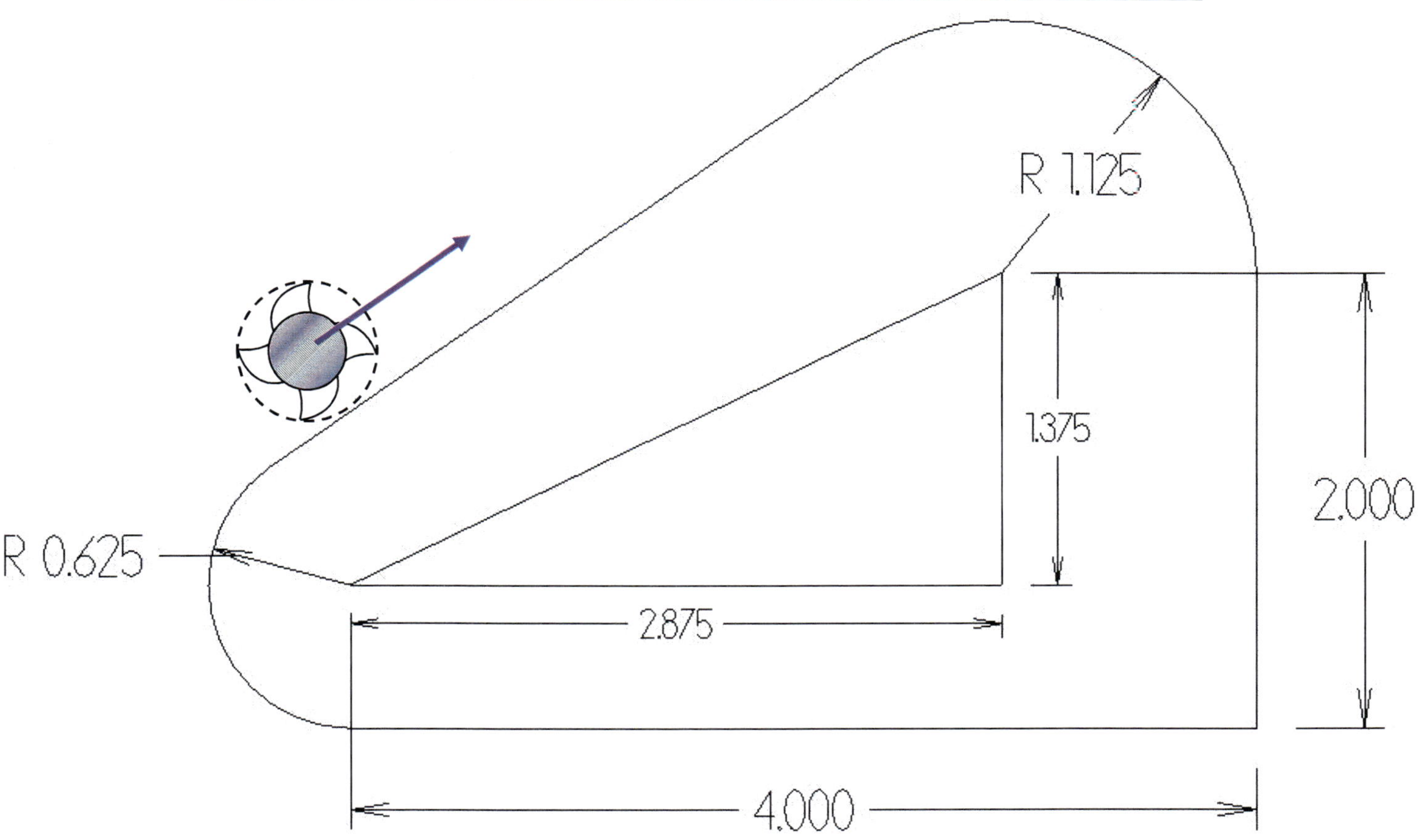

Calculate the angle

$$\theta = \tan^{-1}\left(\frac{\text{opposite}}{\text{adjacent}}\right) + \sin^{-1}\left(\frac{R1 - R2}{\sqrt{A^2 + B^2}}\right)$$

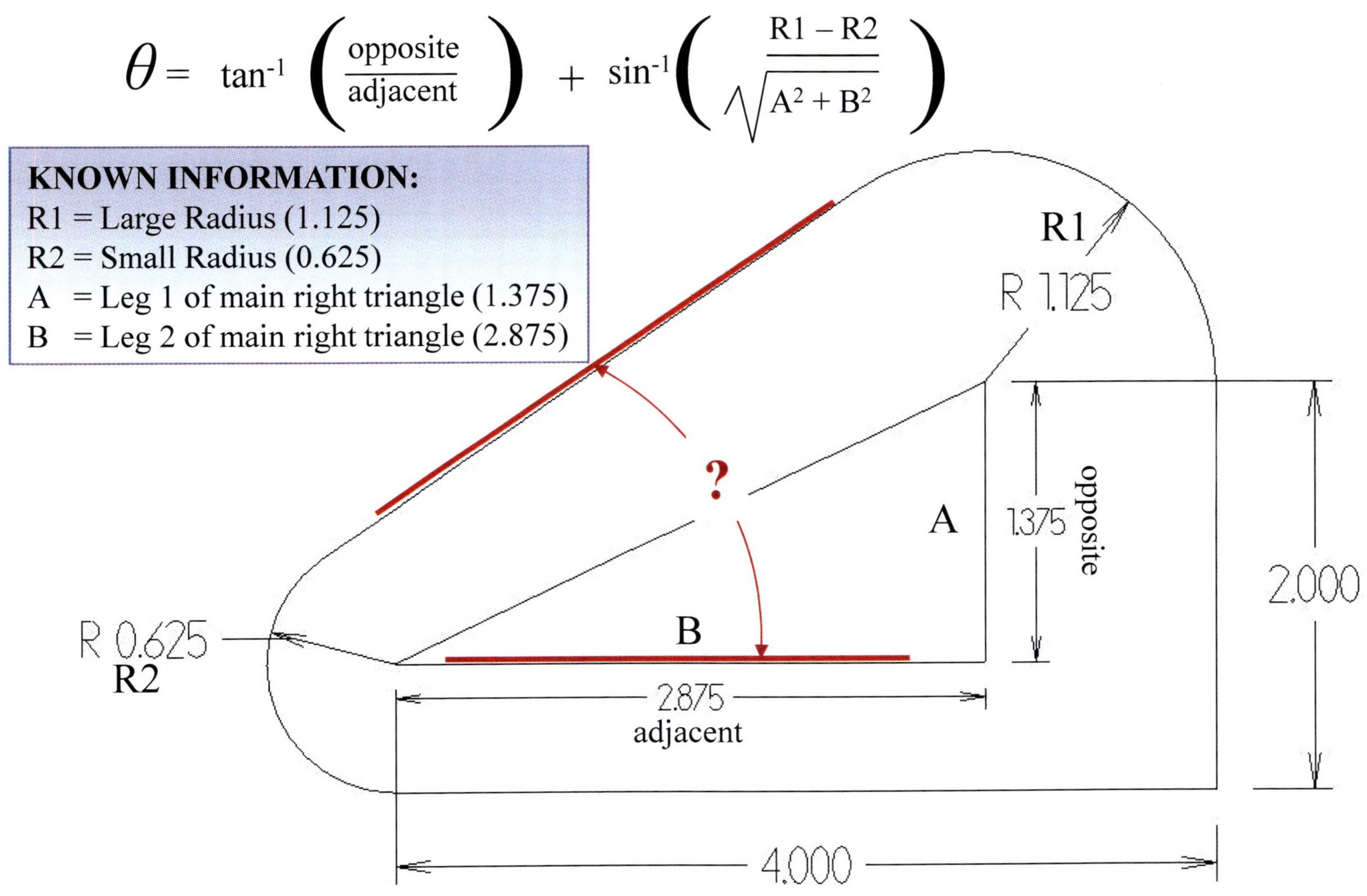

    Advanced CNC Mill Programming and Applied Mathematics Level 2

Step 1: Calculate θ

$$\tan^{-1}\left(\frac{\text{short side}}{\text{long side}}\right) + \sin^{-1}\left(\frac{R1 - R2}{\sqrt{A^2 + B^2}}\right)$$

$$\tan^{-1}\left(\frac{1.375}{2.875}\right) + \sin^{-1}\left(\frac{1.125 - .625}{\sqrt{1.375^2 + 2.875^2}}\right)$$

$$\tan^{-1}\left(.47826\right) + \sin^{-1}\left(\frac{0.5}{\sqrt{10.15625}}\right)$$

$$25.5599 + \sin^{-1}\left(\frac{0.5}{3.18688}\right)$$

$$25.5599 + \sin^{-1} .156893$$

$$25.5599 + 9.026$$

$$\boxed{34.5859}$$

    Advanced CNC Mill Programming and Applied Mathematics Level 2

Step 2a: Calculate triangle 1 using θ and hypotenuse

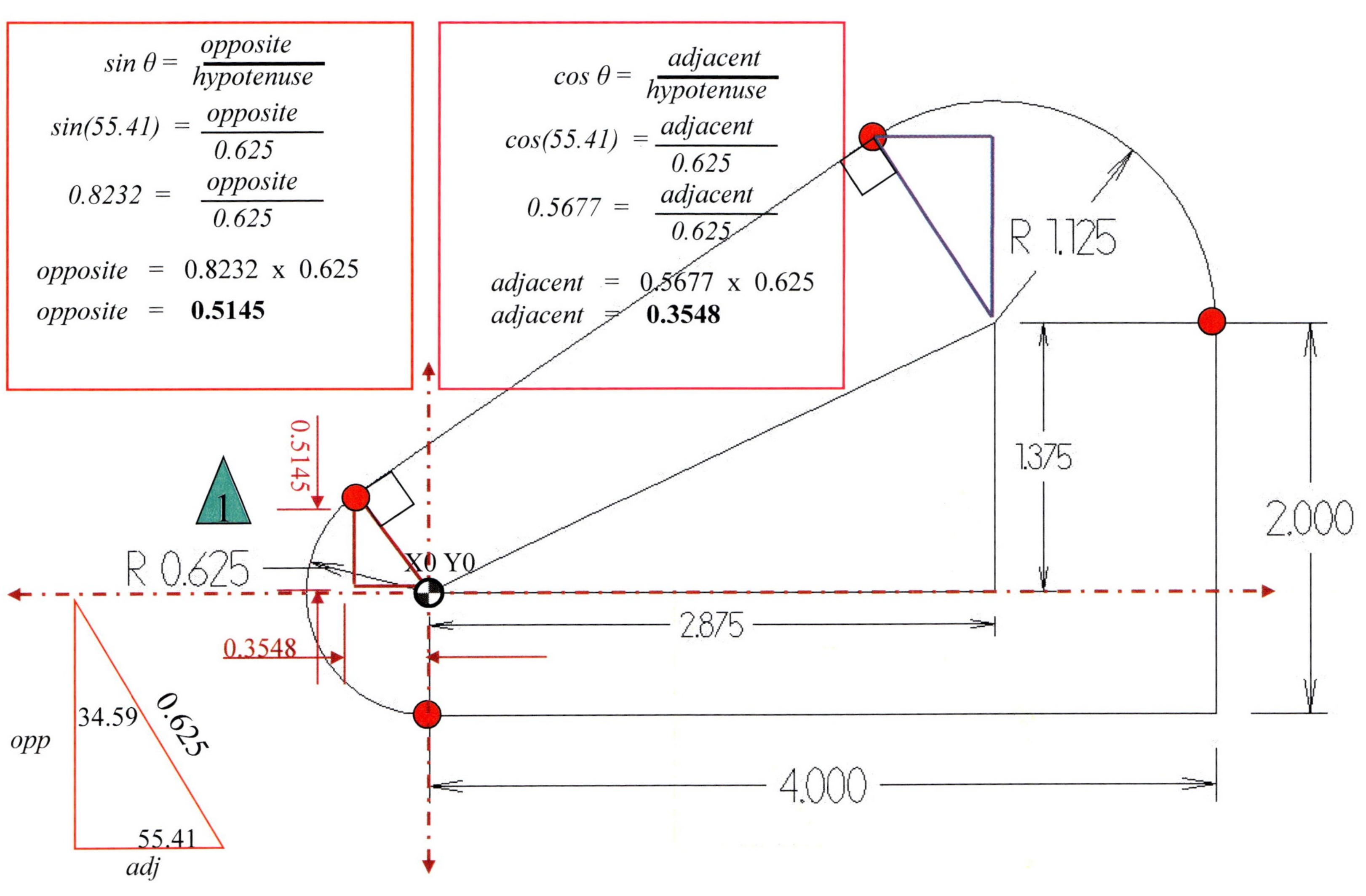

    Advanced CNC Mill Programming and Applied Mathematics Level 2

Step 2b: Calculate triangle 2 using angle and hypotenuse

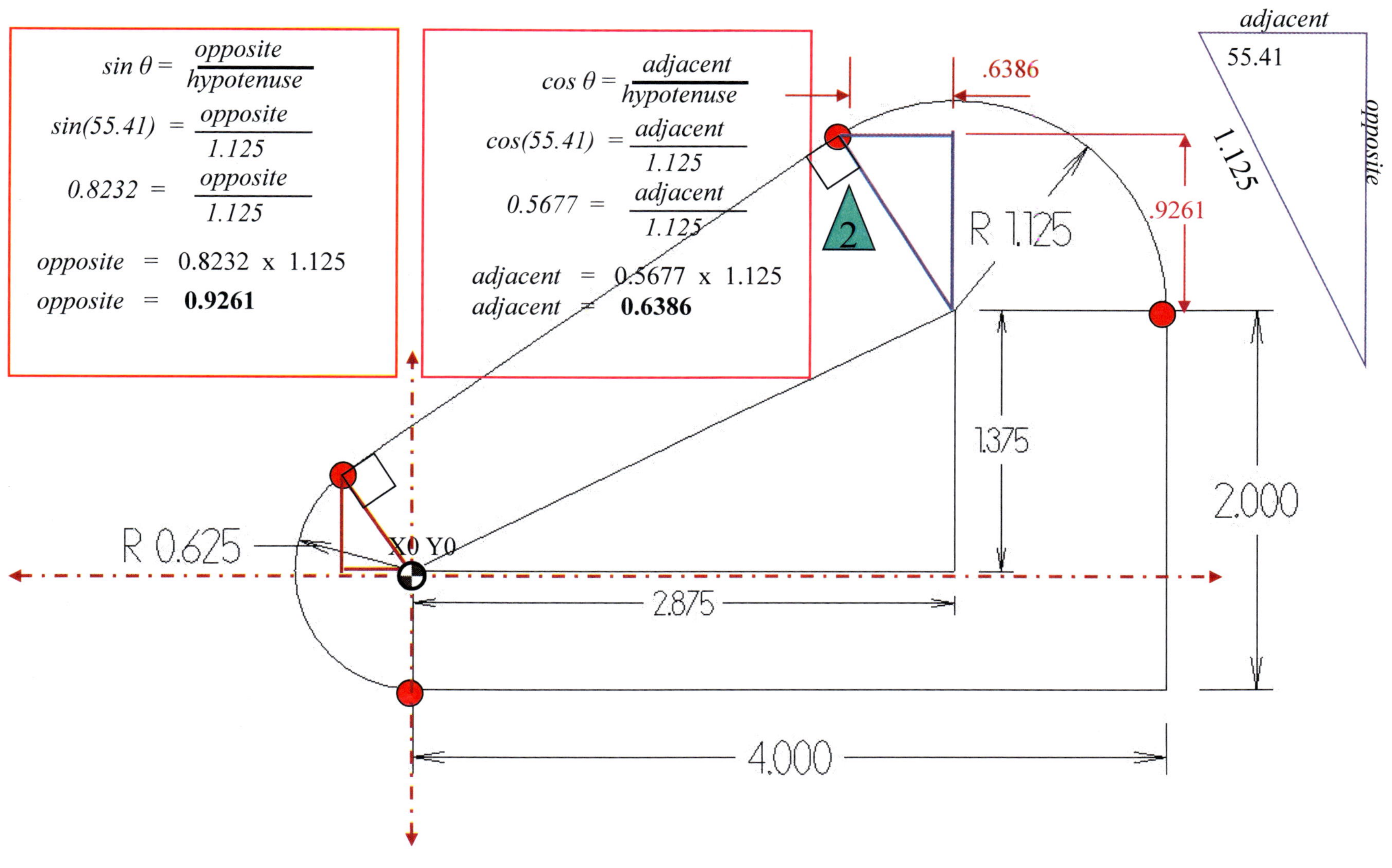

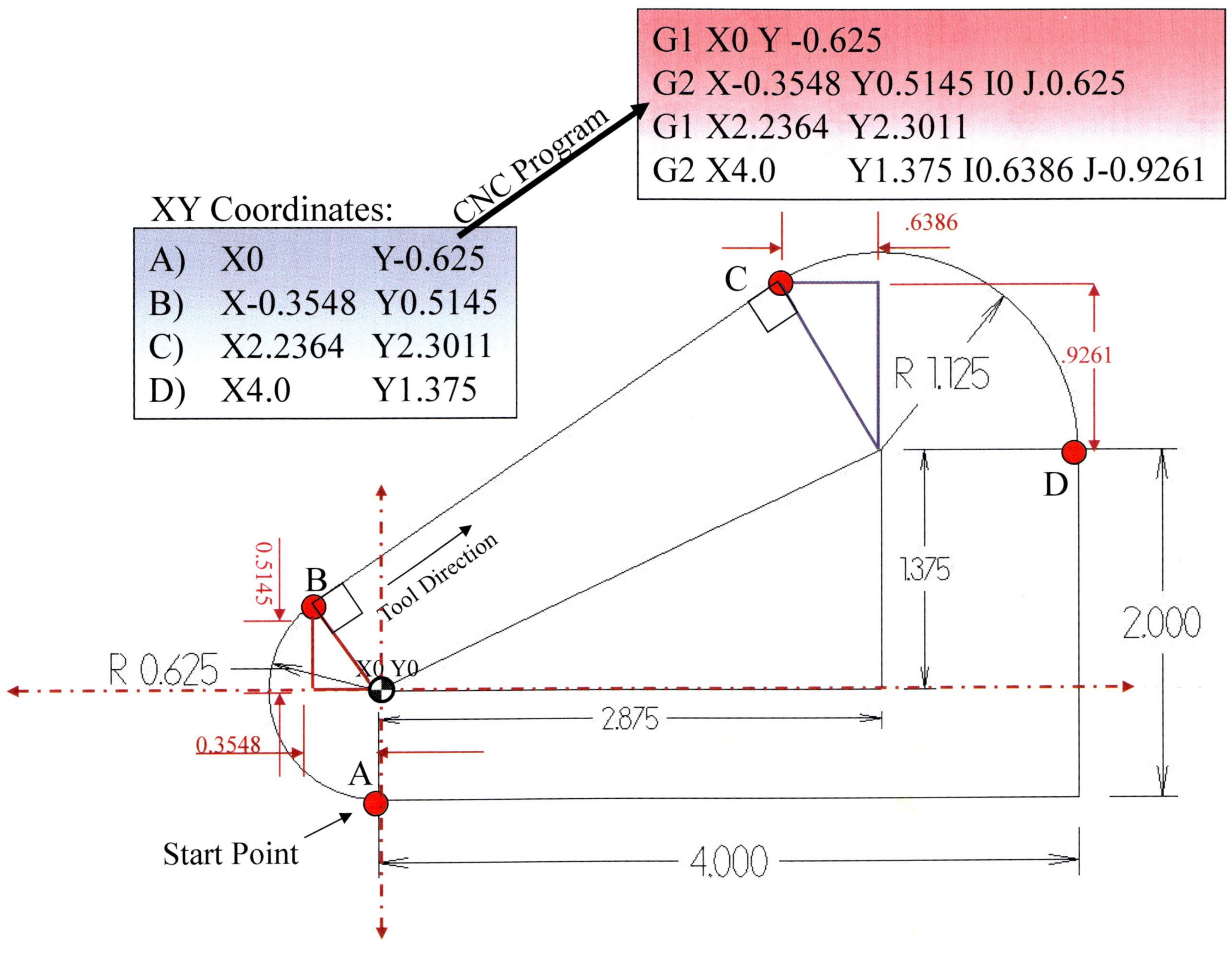

CNC Program
G1 X0 Y -0.625
G2 X-0.3548 Y0.5145 I0 J.0.625
G1 X2.2364 Y2.3011
G2 X4.0   Y1.375 I0.6386 J-0.9261

XY Coordinates:
A)   X0   Y-0.625
B)   X-0.3548 Y0.5145
C)   X2.2364   Y2.3011
D)   X4.0   Y1.375

.6386
C
R 1.125
.9261
D
1.375
2.000
0.5145
B
Tool Direction
R 0.625
X0 Y0
0.3548
A
2.875
Start Point
4.000

CAD Math Solution

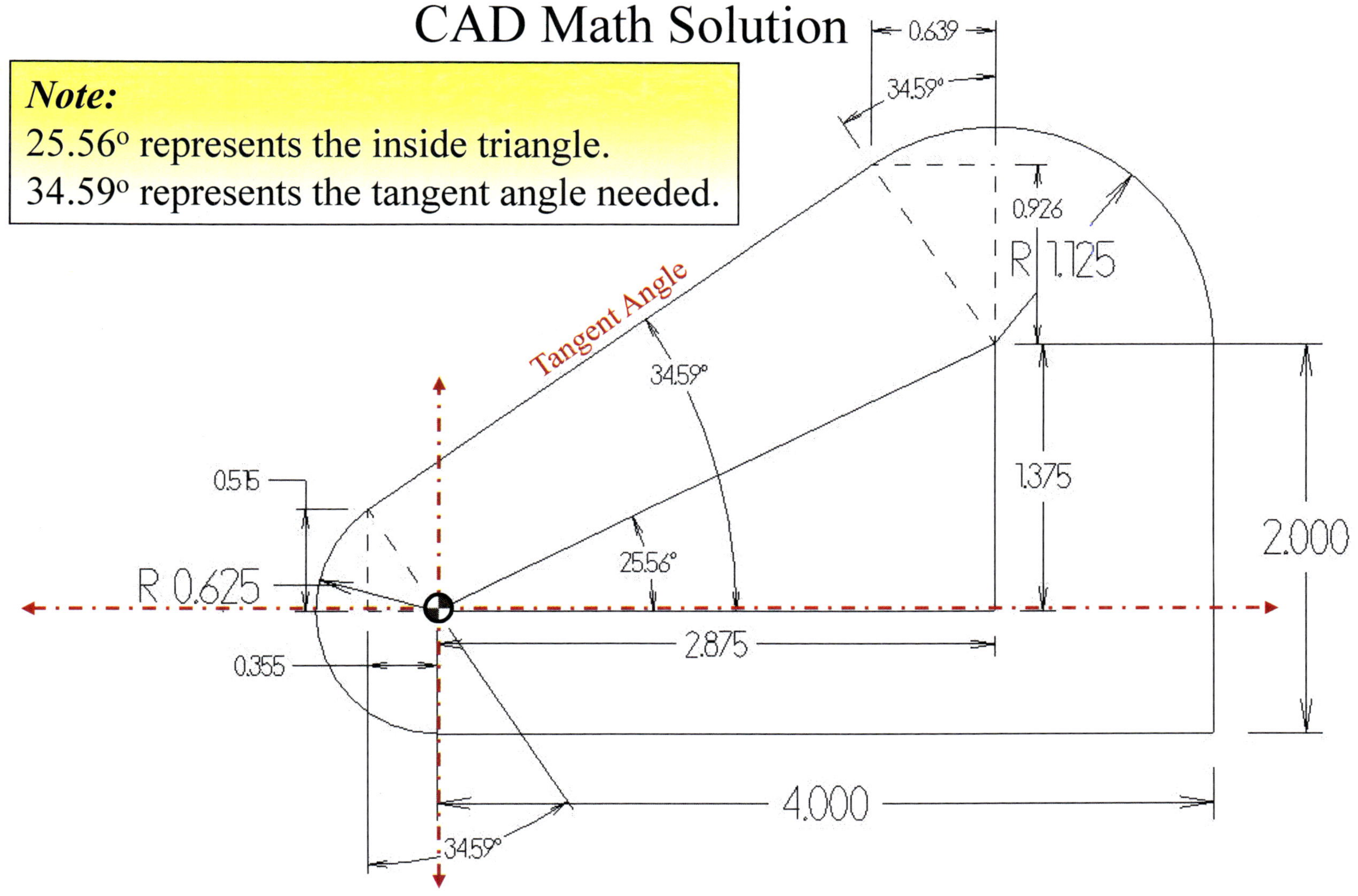

    Advanced CNC Mill Programming and Applied Mathematics Level 2

Mill Project 405

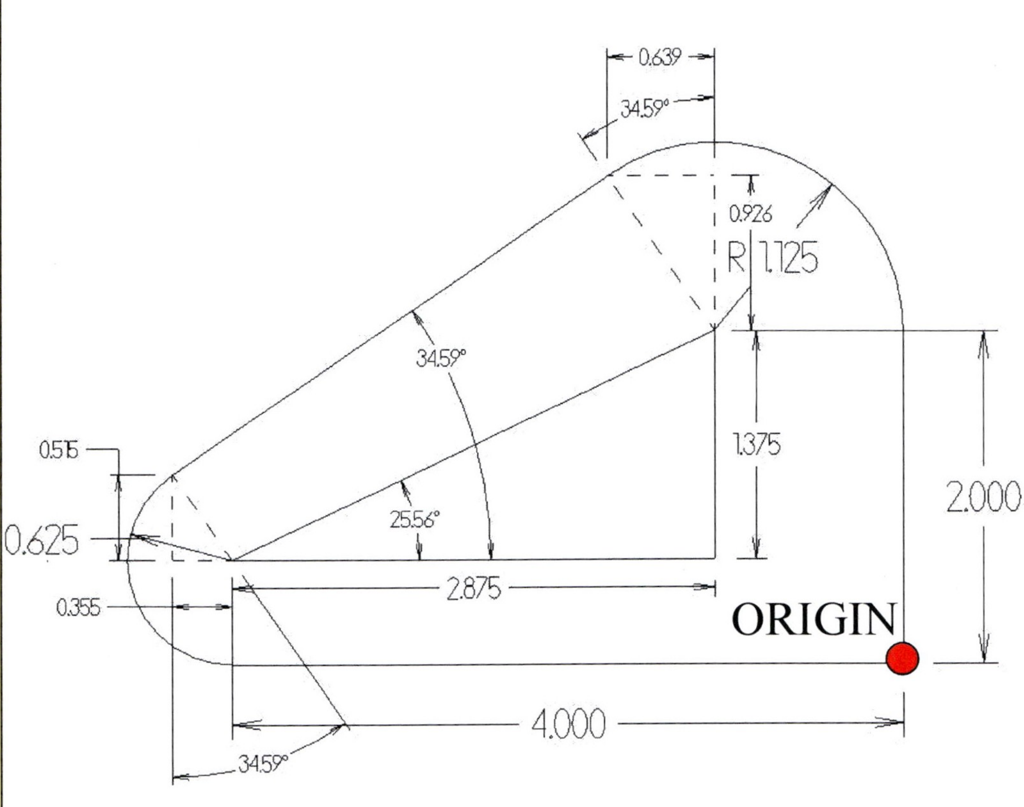


```
%
O405  ( ORIGIN LOWER-RIGHT)
G28 G91 G0 Z0
G54 G90 G0 X1.0 Y-1.0
T2 M6
S1000 M3
G43 H2 Z0.1 M8
G1 Z-0.1 F10.0
G41 D2 Y0
X-4.0
G2 X-4.3548 Y1.1395 I0 J0.625
G1 X-1.7636 Y2.9261
G2 X0 Y2.0 I0.6386 J-0.9261
G1 Y-1.0
G40 X1.0 (turn off CDC)
G28 G91 G0 Z0
M30
%
```


Advanced CNC Mill Programming and Applied Mathematics Level 2

Mill Project 406

Steps:

1. Calculate the angle from circle centers to tangent points.
2. Calculate X & Y adjustments from circle center to tangent points.

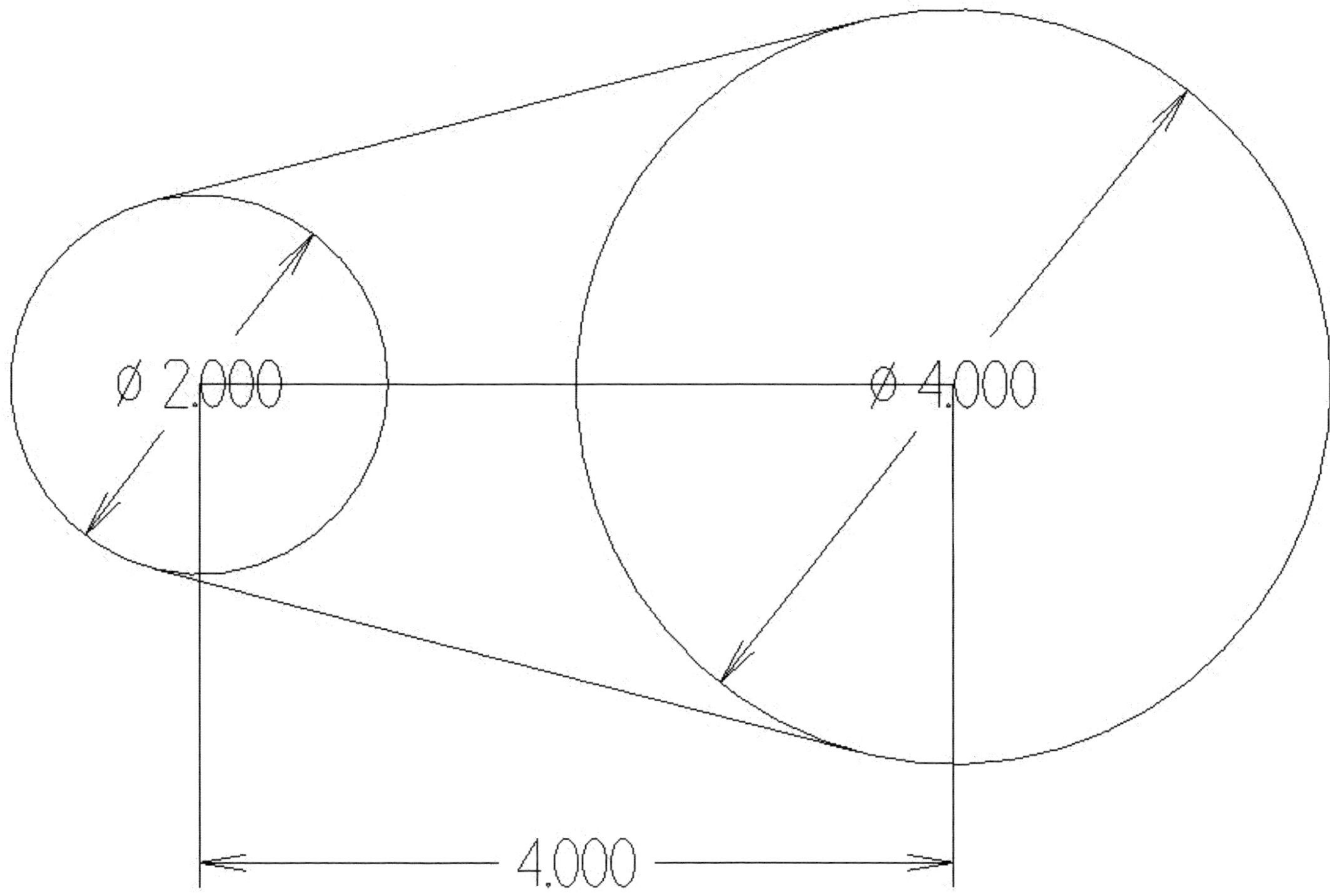

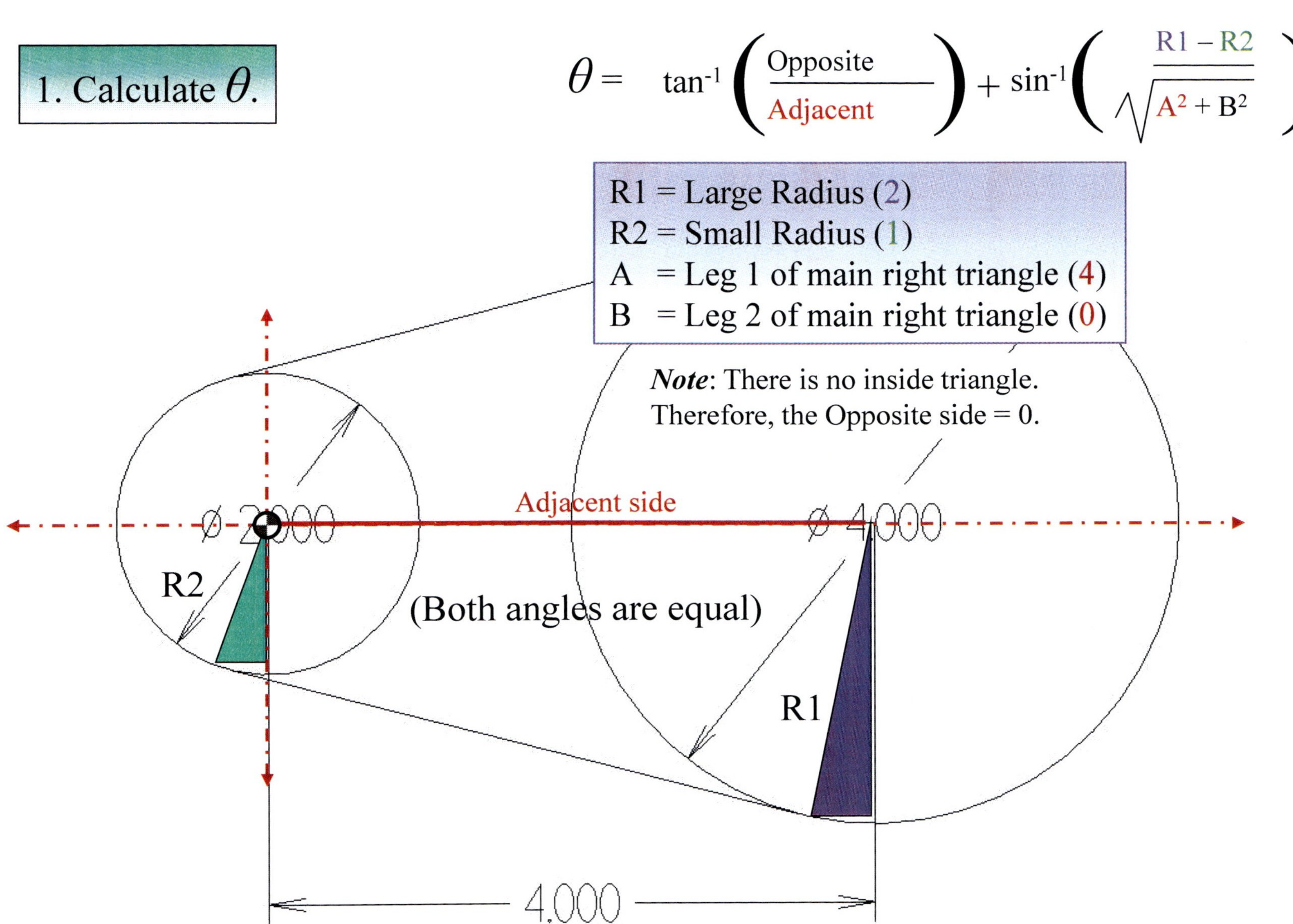

    Advanced CNC Mill Programming and Applied Mathematics Level 2

Step 1: Calculate θ

$$\tan^{-1}\left(\frac{\text{Opposite side}}{\text{Adjacent side}}\right) + \sin^{-1}\left(\frac{R1 - R2}{\sqrt{A^2 + B^2}}\right)$$

$$\tan^{-1}\left(\frac{0}{4}\right) + \sin^{-1}\left(\frac{2 - 1}{\sqrt{4^2 + 0^2}}\right)$$

$$0 + \sin^{-1}\left(\sqrt{\frac{1}{16}}\right)$$

$$0 + \sin^{-1}\left(\frac{1}{4}\right)$$

$$0 + \sin^{-1} .25$$

$$0 + \boxed{14.4775}$$

Note: The first angle = 0 because both circles are on the same horizontal plane.

     Advanced CNC Mill Programming and Applied Mathematics Level 2

$$90 - 14.4775 = 75.5225°$$

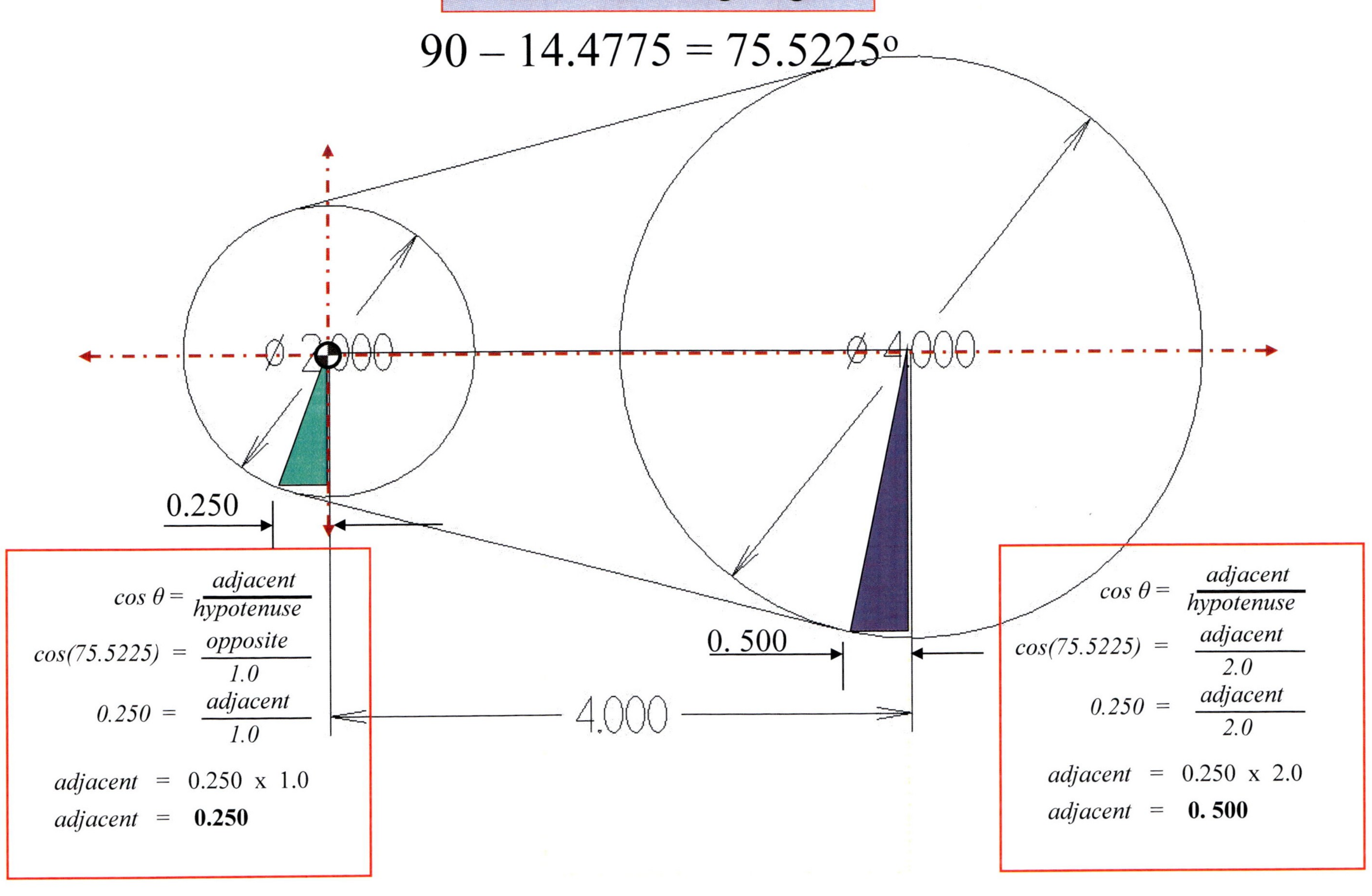

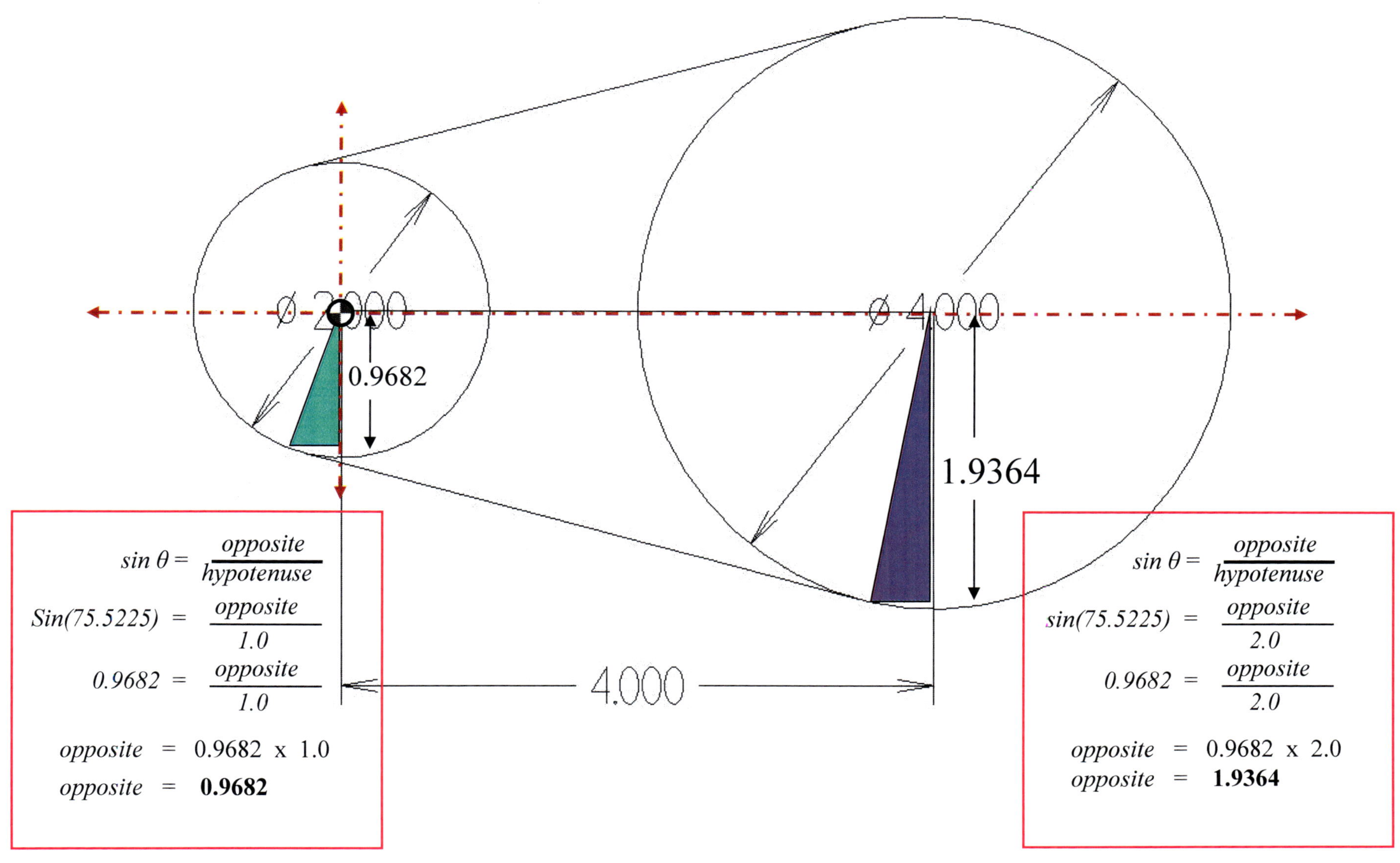

     Advanced CNC Mill Programming and Applied Mathematics Level 2

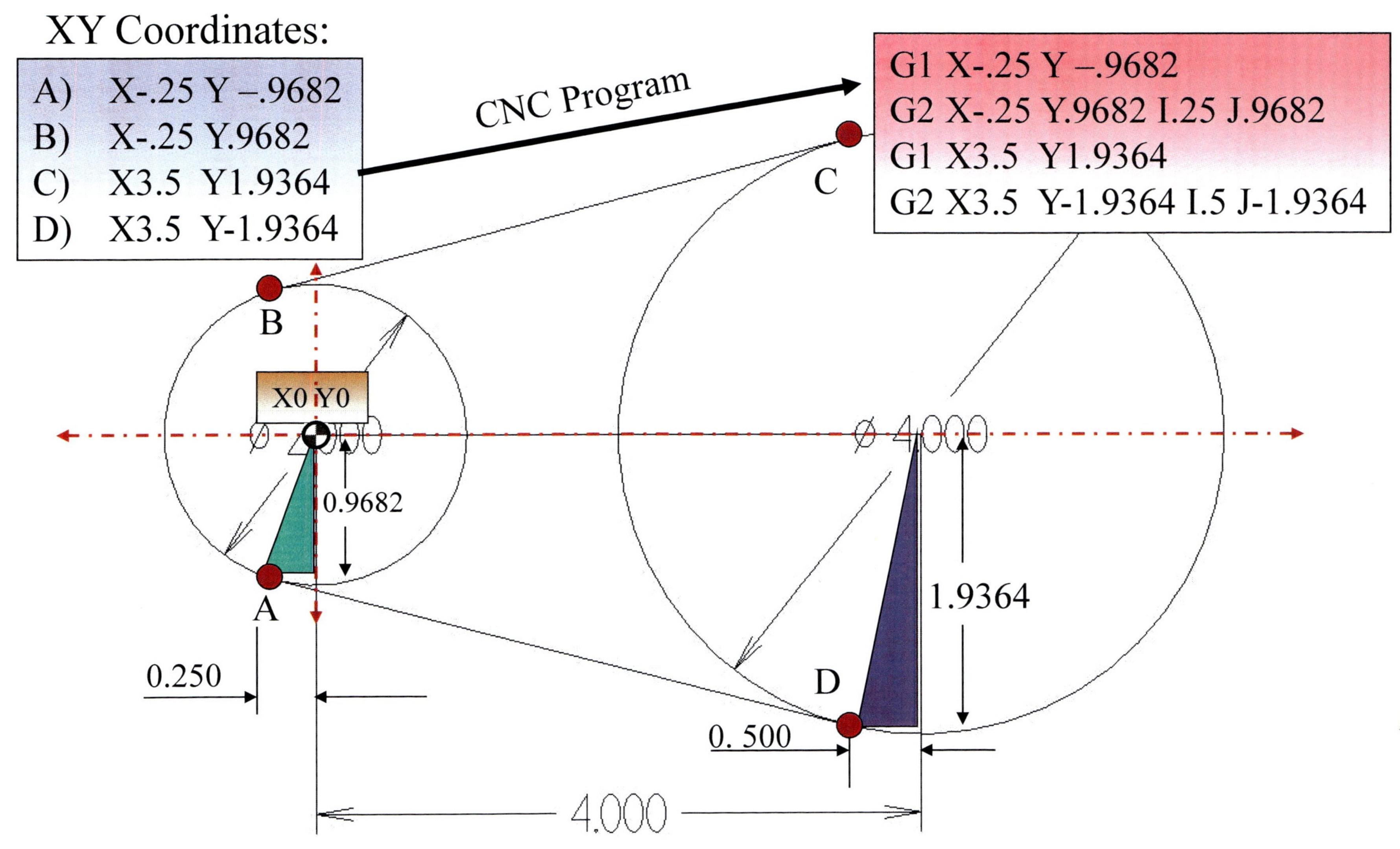

     Advanced CNC Mill Programming and Applied Mathematics Level 2

Finally, calculate the Y-offset necessary to lead-in tangentially. In this case we'll extend the tool 1.0 in X, then calculate the Y-axis offset using right angle trig.

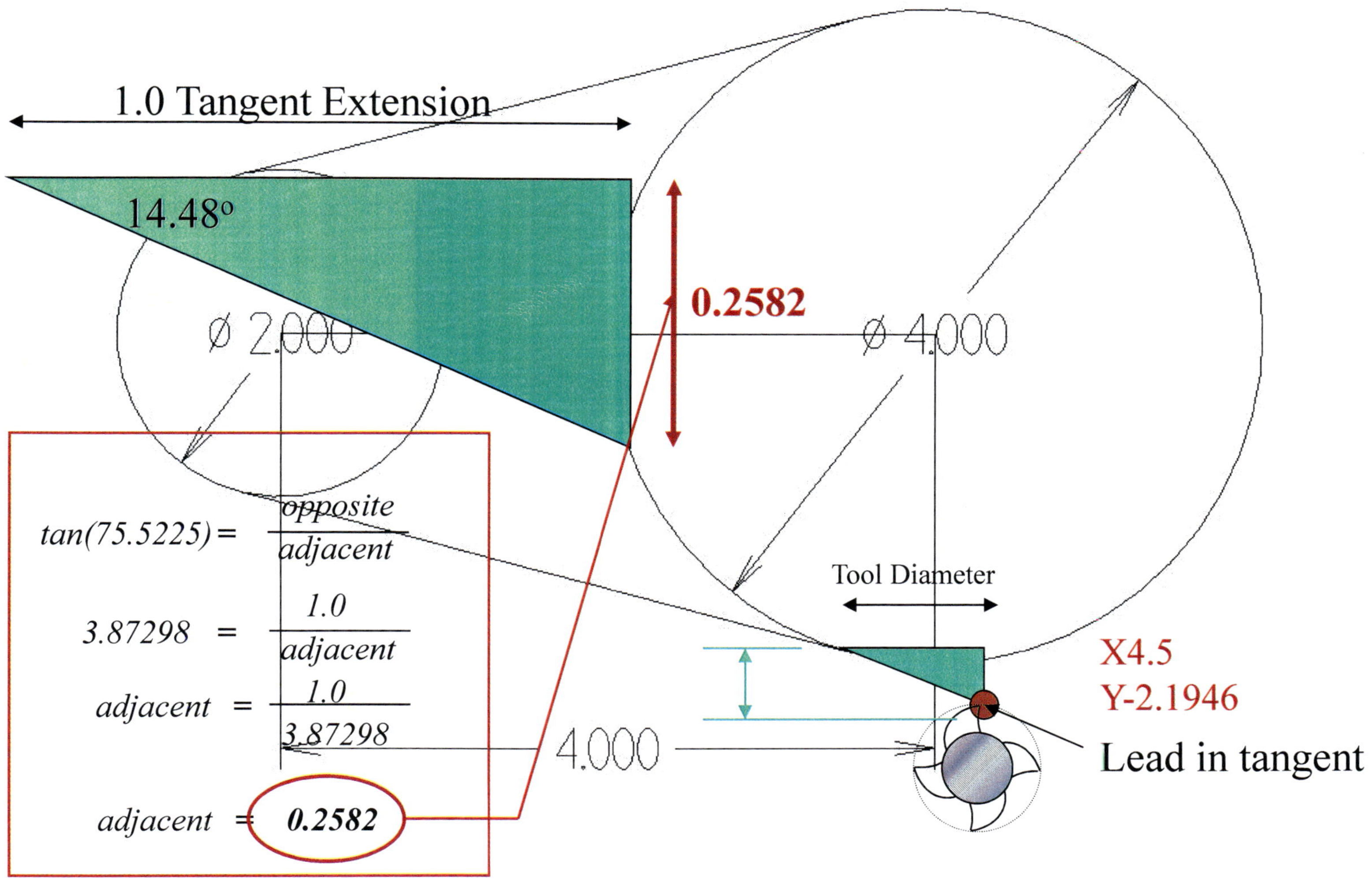

Mill Project 406

```
%
O406  (ORIGIN CENTER OF LEFT CIRCLE)
G28 G91 G0 Z0
G54 G90 G0 X4.5 Y-3.1946
T2 M6
S1000 M3
G43 H2 Z0.1 M8
G1 Z-0.1 F10.0
G41 D2 Y-2.1946
X3.5 Y-1.9346
X-0.250 Y-0.9682
G2 X-0.250 Y0.9682 I0.250 J0.9682
G1 X3.5 Y1.9364
G2 X3.5 Y-1.9364 I0.5 J-1.9364
G3 X3.0 Y-2.4364 R.5 (LEAD OUT WITH .5 ARC)
G40 G1 X4.0 Y-2.9364 (turn off CC)
G28 G91 G0 Z0
M30
%
```

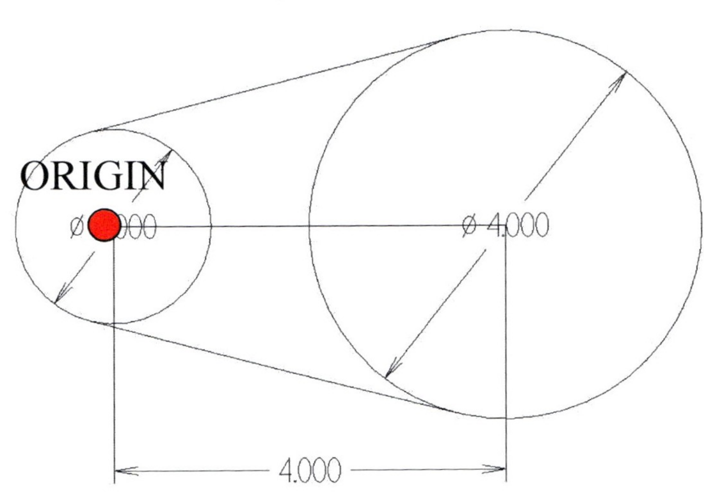

   Advanced CNC Mill Programming and Applied Mathematics Level 2

Answers and Programs

Answers and Programs
Worksheets

Magic Cube Program Exercise (pg. 22)

```
%
O1060 (MAGIC CUBE  1 X 1 X 6.625)
G54 G90 G0 X0 Y0
( ######## HOLES ######## )
T3 M6          ( 3/8 END MILL )
S8000 M3
G90 G0 X0 Y0      ( CENTER OF HOLE )
G43 H3 G0 G90 Z.1 M8
M98 P1070 L5     ( FIVE HOLES )
( ######### CHAMFER ######### )
T16 M6          ( 1/2 CHAMFER TOOL )
S4600 M3
G90 G0 X0 Y0
G43 H16 G0 G90 Z.1 M8
M98 P1080 L5      ( FIVE CHAMFERS )
( ####### CUTOFF SLOTS ######### )
T8 M6          ( 1/8 END MILL )
S8000 M3
G90 G0 X1.125 Y-.6625 ( CENTER OF SLOT  )
G43 H8 G0 G90 Z.1 M8
M98 P1090 L4      ( FOUR SLOTS )
( ###### END OF PROGRAM  ########## )
G90 G0 X0 Y5.     ( OPERATOR LOAD POSITION )
G28 G91 G0 Z0
M30
%
```

```
%
O1070 (MAGIC CUBE SUBPROGRAM 1070
HOLES )

G91 G1 Z-.098  F80.0   (BOTTOM HOLE FIRST)
G1 X-.0313 Y.0313
G3 X-.0313 Y-.0313 R.0313
M97 P100 L15
G3 I.0625 J0 (CLEANUP BOTTOM FACE)
G3 X.0313 Y-.0313 R.0313
G1 X.0313 Y.0313


G90 G0 Z0 (TOP HOLE NEXT)
G91 G1 X-.1093 Y.1093
G3 X-.1093 Y-.1093 R.1093
M97 P200 L6
G3 I.2185 J0 (CLEANUP BOTTOM FACE)
G3 X.1093 Y-.1093 R.1093
G1 X.1093 Y.1093
G90 G0 Z.1


G91 X1.125 (SHIFT TO NEXT HOLE)


N100 (HELICAL INTERP INSIDE HOLE)
G91 G3 I.0625 J0 Z-.02 (4 deg)
M99 (END SUBROUTINE 100)


N200 (HELICAL INTERP LARGER CIRCLE 0.812
DIA )
G91 G3 I.2185 J0 Z-.0325
M99 (end subroutine 200)
M99 (END SUBPRO 1070 HOLES )
```

Answers and Programs
Worksheets

Magic Cube Program Exercise (pg. 22)

```
%
O1080 (MAGIC CUBE SUBPRO 1080 CHAMFER )
( ## 0.005 CHAMFER  ## )
( BOTTOM HOLE 1ST )
(TREAT AS .2 TD; DP = 0.1875 + .1 )
( TREAT AS 0.500 DIAMETER; 0.005 PER SIDE )
G90 G1 Z-.2875 F25.
G91 G1 X-.075 Y.075
G3 X-.075 Y-.075 R.075
G3 I.15 J0
G3 X.075 Y-.075 R.075
G1 X.075 Y.075
( TOP HOLE 2ND )
(TREAT AS .375 TD; DP = 1/2 TD )
( TREAT AS .812 DIAMETER; 0.005 PER SIDE)
G90 Z-.1875  (1/2 TD )
G91 X-.1093 Y.1093
G3 X-.1093 Y-.1093 R.1093
G3 I.2186 J0
G3 X.1093 Y-.1093 R.1093
G1 X.1093 Y.1093
G90 Z.1


G91 X1.125 (SHIFT TO NEXT HOLE)
M99 (END SUBPRO 1080 CHAMFER )
%
```

```
%
O1090 (MAGIC CUBE SUB PRO 1090 SLOTS )

( ## SLOTS ## )

G90 G1 Z-.033 F20.
G1 Y.6625
G0 Z.1
G0 Y-.6625
G1 Z-.066
G1 Y.6625
G0 Z.1
G0 Y-.6625
G1 Z-.1
G1 Y.6625
G0 Z.1


G91 X1.125 (SHIFT TO NEXT HOLE)

M99 ( END SUB PRO 1090 SLOTS )
%
```

Answers and Programs
Projects

Mill Project 301 (pg. 101)

```
%
O301 ( ORIGIN CENTER )
G54 G90 G0 X0 Y0
T3 M6  (90 DEG SPOT DRILL)
S1000 M3
G43 H3 Z0.1 M8

( ## QUADRANT 1 ## )
G81 X1.2703 Y0.5262 R0.1 Z-0.1975 F10.
X0.9723 Y0.9723
X0.5262 Y1.2703

( ## QUADRANT 2 ## )
X-0.5262 Y1.2703
X-0.9723 Y0.9723
X-1.2703 Y0.5262
( ## QUADRANT 3 ## )
X-1.2703 Y-0.5262
X-0.9723 Y-0.9723
X-0.5262 Y-1.2703

( ## QUADRANT 4 ## )
X0.5262 Y-1.2703
X0.9723 Y-0.9723
X1.2703 Y-0.5262
G80 M9

G28 G91 G0 Z0
M30
%
```

Mill Project 302 (pg. 105)

```
%
O302 ( ORIGIN CENTER )
      ( TOOL ON )
G54 G90 G0 G80 G40 X-4.0 Y-3.0
T2 M6
S1000 M3
G43 Z0.1
G1 H2 Z-0.1 F10. M8
G41 D2 X-3.0
Y1.0
X-2.4226 Y2.0 (CHAMFER 1)
X-1.0
Y1.0
X1.0
Y2.0
X2.4226
X3.0 Y1.0
Y-1.0
X2.4226 Y-2.0
G1 X1.0
Y-1.0
X-1.0
Y-2.0
X-2.4226
X-3.0 Y-1.0
X-4.0
G40 Y-3.0
G28 G91 G0 Z0
M30
%
```


   Advanced CNC Mill Programming and Applied Mathematics Level 2

Answers and Programs
Projects

Mill Project 303 (pg. 108)

```
%
O303 ( ORIGIN LOWER-LEFT )
      ( TOOL ON )
G54 G90 G0 X-1.0 Y-1.0
T2 M6
S1000 M3
G43 H2 Z0.1
G1 Z-0.1 F10.0 M8
G41 D2 X0
Y1.9142
X2.0858 Y4.0
X4.0
Y3.0
X2.5
X1.0 Y1.5
Y0
X-1.0
G40 Y-1.0
G28 G91 G0 Z0
M30
%
```

Mill Project 304 (pg. 111)

```
%
O304 ( ORIGIN CENTER )
      ( TOOL ON )
G54 G90 G0 X2.0 Y-2.5 (START IN LR)
T2 M6
S1000 M3
G43 H2 Z0.1
G1 Z-0.1 F10.0 M8
G41 D2 Y-1.5
X-0.866
X-1.732 Y0
X-.866 Y1.5
X0.866
X1.732 Y0
X0.866 Y-1.5
Y-2.5
G40 X2.0
G28 G91 G0 Z0
M30
%
```

Answers and Programs
Projects

Mill Project 305 1 (pg. 115)

```
%
O3051 (Male Dovetail; Stock is 4" Deep)

(Safe start and prepare for Cutter comp)
G54 G90 G0 G40 G80 X-0.6204 Y-1.5   (Pt
1)
T15 M6
S1200 M3
G43 H15 G90 G0 Z0
G90 G1 Z-0.1042 F20. (Feed to 1st Z step)
G41 D15 Y0  F20. (Feed to compensated
position)
Y4.1                (Clear stock by 0.1)
X0.6204               (Move to Pt 2)
Y-0.1               (Clear stock by 0.1)
Z-0.2084              (Feed to 2nd Z step)
X-0.5602              (Move to Pt 3)
Y4.1
X0.5602              (Move to Pt 4)
Y-0.1
Z-0.3125              (Feed to Final Z step)
X-0.500              (Move to Pt 5)
Y4.1
X0.500               (Move to Pt 6)
Y-0.1
G28 G91 G0 Z0
M30
%
```

Mill Project 305 2 (pg. 117)

```
%
O3052 (Female Dovetail; Stock is 4" Deep)

(Safe start and prepare for Cutter comp)
G54 G90 G0 G40 G80 X0.6227 Y-1.5  (Pt 1)
T15 M6
S1200 M3
G43 H15 G90 G0 Z0
G90 G1 Z-0.1042 F20. (Feed to 1st Z step)
G41 D15 Y0 F20. (Feed to compensated position)
Y4.1             (Clear stock by 0.1)
X-0.6227         (Move to Pt 2)
Y-0.1            (Clear stock by 0.1)
Z-0.2084           (Feed to 2nd Z step)
X0.6829            (Move to Pt 3)
Y4.1
X-0.6829           (Move to Pt 4)
Y-0.1
Z-0.3125           (Feed to Final Z step)
X0.7431            (Move to Pt 5)
Y4.1
X-0.7431           (Move to Pt 6)
Y-0.1
G28 G91 G0 Z0
M30
%
```

Answers and Programs
Projects

Mill Project 401 1 (pg. 120)

```
%
O4011  ( ORIGIN RIGHT-CENTER  )
       ( PROFILE 1 of 2 )
G28 G91 G0 Z0
G54 G90 G0 X0.75 Y0
T10 M6
S1000 M3
G43 H10 Z0.1 M8
G1 Z-0.1 F10.0
G41 D10 Y-0.38 (TURN ON CC)
X-.13                         (B)
G2 X-0.13 Y0.38 I0 J0.38     (C)
G1 X0.75
G40 Y0 (TURN OFF CC)
G28 G91 G0 Z0
M30
%
```

Mill Project 401 2 (pg. 123)

```
%
O4012  ( ORIGIN RIGHT-CENTER )
       ( PROFILE 2 of 2 )
G28 G91 G0 Z0
G54 G90 G0 X0.75 Y0
T10 M6
S1000 M3
G43 H10 Z0.1 M8
G1 Z-0.1 F10.0
G41 D10 Y0.69          (TURN ON CC)
X-0.13                              (B)
G3 X-0.7653 Y0.2693 I0 J-0.69       (C)
G2 X-0.885 Y0.19 I-0.1197 J0.0507   (D)
G1 X-1.06                           (E)
G2 X-1.19 Y0.32 I0 J0.13            (F)
G1 Y0.585                           (G)
G3 X-2.19 Y0.585 I-0.5 J0           (H)
G1 Y-0.585                          (I)
G3 X-1.19 Y-0.585 I0.5 J0           (J)
G1 Y-0.32                           (K)
G2 X-1.06 Y-0.19 I0.13 J0           (L)
G1 X-0.885                          (M)
G2 X-0.7653 Y-0.2693 I0 J-0.13      (N)
G3 X-0.13 Y-0.69 R.69               (O)
G1 X0.75                            (P)
G40 Y0               (TURN OFF CC)
G28 G91 G0 Z0
M30
%
```

Answers and Programs
Projects

Mill Project 402 (pg. 124)

```
%
O402 ( Outer profile )
G54 G90 G0 X0 Y-2.8738 (Move to Home base
position)
T2 M6
S1000 M3
G43 Z0.1
G1 H2 Z-0.1 F10. M8
G41 D2 X0.5 Y-2.3738        (1ST Base)
G3 X0 Y-1.8738 R0.5         (Arc into 2ND base)
G3 X-2.9413 Y-2.8305 R5.0
G2 X-2.9413 Y2.8305 I-2.0587 J2.8305
G3 X2.9413 Y2.8305 R5.0
G2 X2.9413 Y-2.8305 I2.0587 J-2.8305
G3 X0 Y-1.8738 R5.0         (2nd Base)
G3 X-0.5 Y-2.3738 R0.5      (Arc off to 3rd Base)
G40 G1 X0 Y-2.8738
G28 G91 G0 Z0
M30
%
```

Mill Project 403 (pg. 129)

```
%
O403 ( ORIGIN CENTER  TOOL ON )
G54 G90 G0 X0 Y-3.0
T2 M6
S1000 M3
G43 H2 Z0.1
G1 Z-.1 F10.0 M8
G41 D2 X0.5 Y-2.5 (1ST BASE)
G3 X0 Y-2.0 R0.5 (2ND BASE)
G1 X-2.0
G2 X-2.866 Y-0.5 R1.0 (LL)
G1 X-1.7113 Y1.5
G2 X-0.8453 Y2.0 R1.0 (UR)
G1 X2.0
G2 X2.866 Y0.5 R1.0 (UR)
G1 X1.7113 Y-1.5
G2 X0.8453 Y-2.0 R1.0 (LR)
G1 X0
G3 X-0.5 Y-2.5 R0.5 (ARC OFF)
G40 G1 X0 Y-3.0
G28 G91 G0 Z0
M30
%
```


    Advanced CNC Mill Programming and Applied Mathematics Level 2

Answers and Programs
Projects

Mill Project 404 (pg. 133)

```
%
O404  ( ORIGIN CENTER)
G28 G91 G0 Z0
G54 G90 G0 X55. Y-45.5 (Start of baseball
diamond)
T6 M6 (3/8 END MILL)
S1000 M3
G43 H6 Z2.54 M8
G1 Z-2.54 F100.
G41 D6 X75.0 Y-25.5 (1ST BASE)
G3 X55.0 Y-5.5 R20. (2ND BASE)
G1 X34.82
G3 X8.84 Y-13.33 I0 J-47.0
G2 X8.84 Y13.33 I-8.84 J13.33
G3 X34.82 Y5.5 I25.98 J39.17 (Completes left
end)
G1 X83.38 Y5.5
G3 X99.08 Y11.92 R23.0
G2 X99.08 Y-11.92 R-16. (ARC > 180 NEEDS -)
G3 X83.38 Y-5.5 R23.0
G1 X55.0 Y-5.5          (2nd BASE)
G3 X35.0 Y-25.5 R20.0   (3rd BASE)
G1 G40 X55.0 Y-45.5     (Home Base-Turn off
CC)
G28 G91 G0 Z0
M30
%
```

Mill Project 405 (pg. 163)

```
%
O405  ( ORIGIN LOWER-RIGHT)
G28 G91 G0 Z0
G54 G90 G0 X1.0 Y-1.0
T2 M6
S1000 M3
G43 H2 Z0.1 M8
G1 Z-0.1 F10.0
G41 D2 Y0
X-4.0
G2 X-4.3548 Y1.1395 I0 J0.625
G1 X-1.7636 Y2.9261
G2 X0 Y2.0 I0.6386 J-0.9261
G1 Y-1.0
G40 X1.0 (turn off CC)
G28 G91 G0 Z0
M30
%
```

Answers and Programs
Projects

Mill Project 406 (pg. 171)

```
%
O406  (ORIGIN CENTER OF LEFT CIRCLE)
G28 G91 G0 Z0
G54 G90 G0 X4.5 Y-3.1946
T2 M6
S1000 M3
G43 H2 Z0.1 M8
G1 Z-0.1 F10.0
G41 D2 Y-2.1946
X3.5 Y-1.9346
X-0.250 Y-0.9682
G2 X-0.250 Y0.9682 I0.250 J0.9682
G1 X3.5 Y1.9364
G2 X3.5 Y-1.9364 I0.5 J-1.9364
G3 X3.0 Y-2.4364 R.5 (LEAD OUT WITH .5
ARC)
G40 G1 X4.0 Y-2.9364 (turn off CC)
G28 G91 G0 Z0
M30
%
```